Digital Transformation Success Secrets!

The Ultimate Guide to Business, Career & Life Success

Marty Fox

Copyright © 2018 All Rights Reserved by Fox Technology Ventures, LLC

No part of this work may be reproduced or transmitted in any form or by any means without express written consent of the publisher.

Neither the author nor the publisher of this book is engaged in rendering, by the sale of this book, medical, legal, accounting or any other professional services. The reader is encouraged to employ the services of a competent professional in such matters.

Printed in the United States of America

Dedication

This book is dedicated to my wife Debbie, son David and all of my family and friends who put up with the hundreds of weekend and evening hours I devoted to creating this book. Thank you for your unconditional loving support.

This book is also dedicated to you, the next digital success story! Work hard and never ever give up. You can change the world.

Acknowledgments

A very special thank you to my son David and brother Joe for their valuable input in shaping this book. Their ideas and feedback brought the book to a lofty and unique level.

This book would not have been possible without the professional and tireless editing and input of my wife, Debbie. She transformed my ideas and experiences into a product we are proud to share with the world.

I have been very fortunate for the hundreds of close friendships I have been privileged to make during a very rewarding career. So many wonderful people have helped me become successful as a digital technologist, entrepreneur and executive. It would take a book to name everyone that impacted my life.

In the making of this book I would like to give thanks to the many organizations and individuals that were kind and sharing during email and phone conversations.

Thank you all for your friendship and professionalism!

Special FREE Bonus
Digital Success Newsletter

Digital transformation and disruption are quickly changing the world. It is important for you to understand and stay ahead of change.

As a free bonus to my book readers I would like to continue providing you with value after the book with a free subscription to my Digital Transformation Success Secrets Newsletter. Change does not stop and neither can you.

If you do not enjoy the newsletter you can cancel at any time.

Each issue is packed with:
- Late breaking digital business and career alerts, tips and techniques
- Digital software and hardware recommendations, reviews, free trials and product discounts
- New career success opportunities, success habits and possible pitfalls
- A comprehensive online listing and links to ALL resources in this book and many future digital companies and products as I discover them

To begin your free subscription today visit DigitalSuccessBook.com/newsletterbonus

Disclaimer

This book is presented solely for educational and entertainment purposes. The author and publisher are not offering it as legal, accounting, or any other professional services advice. While best efforts have been used in preparing this book, the author and publisher make no representations or warranties of any kind and assume no liabilities of any kind with respect to the accuracy or completeness of the contents and specifically disclaim any implied warranties of merchantability or fitness of use for a particular purpose.

Neither the author nor the publisher shall be held liable or responsible to any person or entity with respect to any loss or incidental or consequential damages caused, or alleged to have been caused, directly or indirectly, by the information or programs contained herein. No warranty may be created or extended by sales representatives or written sales materials.

Every company is different and the advice and strategies contained herein may not be suitable for your situation. You should seek the services of a competent professional before beginning any improvement program. The ideas in this book worked for me. You should be mindful of what works for you and what does not work in your program or life. Any likeness to actual persons, either living or dead, is strictly coincidental.

Neither the author or anyone associated with this book warrants the accuracy, reliability or timeliness of any information published in the book, nor endorses any products or services linked from the book, and shall not be held liable for any losses caused by reliance on the accuracy, reliability or timeliness of such information. Portions of the information may be incorrect or not current. Any person or entity that relies on any information obtained from this book does so at his or her own risk.

All product and company names are trademarks™ or registered ® trademarks of their respective holders. Use of them does not imply any affiliation with or endorsement by them.

The views and opinions expressed in the book are those of the author and do not reflect the official policy or position of any company or agency of the U.S. government.

Let's Get Digital!

Welcome! Writing 'Digital Transformation Success Secrets' was a labor of love. I am confident you will enjoy the exciting digital journey we are about to embark on.

Your timing is perfect! We are firmly in an age where every business, career and facet of our personal life is impacted by digital transformation. Whether you are a business owner, executive, employee, career seeker, student or you just want to learn more about how digital can transform your life, this is the book for you.

You do not need any experience. I will quickly help you become digital savvy by sharing practical easy-to-understand hands-on experiences, tips and techniques I use every day. I will also share the best practices and inspirational stories from other successful digital entrepreneurs and executives.

This is not a book consisting of 'pie in the sky' overly complex theoretical frameworks that take years to implement. This is a hands-on book about what has worked for me and others in the real-world.

This book is as much about ideas, people and process as it is about technology. The greatest technology in the world will go nowhere without ideas, innovators, creators, entrepreneurs, executives and employees who are not afraid to boldly take on the world and make it a better place.

Many ideas I will share with you are unique, ultra-innovative and different. These 'moonshots' have often led to huge payoffs for me as an entrepreneur and corporate leader. Taking 'the road less traveled' has been key to my digital success.

Throughout the book, I feature 175 amazing digital products and innovative companies you need to know about. They have cracked the digital code and are reinventing a better world. Many of the solutions will have you shaking your head and thinking, 'that is just not possible', but it is and I include details, stories and links for further information.

Executives, business owners and entrepreneurs - Whether you are a leader of a global organization or are launching a one-person start-up, this is the book for you. I have succeeded on every level. You will discover how to reinvent your business using digital mindset, technology and process. You will learn how to be a disruptor, how to prevent being disrupted and how to find hidden digital opportunities and threats.

Employees, job-seekers and students - If you are interested in building your career, this is the book for you. You will learn digital success habits, tips and techniques that will make you more valuable and employable. Being digitally savvy will provide you with clues to the future. You will become more confident and secure in your abilities.

In the digital age, every employee is responsible for taking ownership and helping their employer grow. I can assure you, from personal career success, that understanding the 'digital difference' will enable you to be that highly valued employee. This is the book that will help you quickly become one of the most digital savvy people in your organization. I promise you that one great digital transformation idea can make you a superstar and supercharge your career.

The book reveals 4 steps to digital business and career success:

Step 1 - You will learn how to develop a proven digital mindset and prepare for success. You will learn about transformation, convergence and why now is the perfect time to get digital. You will learn 22 habits for digital success and 7 habits that will lead to failure. I will show you how to discover hidden value at your fingertips. You will learn of digital tools that can magically transform torrents of raw data into insights. I will share how to 'digitally play' for business and career success.

Step 2 - You will learn about the amazing new digital technologies that are changing the world and how to apply them in straightforward terms. I will demystify each technology in simple terms and provide many examples of how you can creatively apply them to your business, career or personal life.

Step 3 - You will learn how innovation and combinations of the new technologies are being used to reinvent business and radically transform lives. Every chapter in this section includes real-world examples of valuable digital products and the companies that created them.

Step 4 - You will learn how to build digital greatness using idea generating techniques, best practices, digital strategies, new digital business models, proper planning, execution and testing. You will learn what works and what does not work in the real-world. I will share my digital success secrets plus that of other successful digital entrepreneurs and executives.

For example:
- A laid-off widowed grandmother was frustrated while searching the web for a peach-pie recipe. Hundreds of porn sites with 'Peaches' appeared, which is not what she was looking for. She had an idea to build a safe and friendly digital platform where people could search and socialize. Somehow, she built it, grew it and was profiled in hundreds of media outlets globally, including a full-page in People Magazine and 3/4 of a page in the New York Times Sunday Business Section. She created new value for thousands of people worldwide and became so successful her digital property and brand was acquired by a larger company and she happily worked with them for years.
- A sustainable entrepreneur launched a company harvesting worm-poop as a college student. He hired a brilliant engineer and they are now making the world a safer, healthier and cleaner place. They take on the biggest challenges and are admired world-wide for their environmental success and business success.
- Two technologists with full time careers crafted a 'business plan' on the back of a napkin during a burger and fries lunch. They quickly filled a global need and transformed their idea into a leading digital social media destination that delights people in over 100 countries.

I would love to hear from you. If you have questions or feedback please contact me.
Together we will win the Digital War!
Marty Fox
Email:Mfox@DigitalSuccessBook.com

Table of Contents

Step 1 - Developing the Foundation for Digital Success 1

Rocking the World with Digital 2
22 Habits for Digital Success 14
7 Habits for Digital Failure 22
Discovering Extraordinary Value in Experimentation 25
Do You Know What Business You are 'REALLY' In? 30
How to Disrupt the Disruptors 35
How to Digitally Play, Prototype and Profit 43

Step 2 - Understanding the New Technologies 53

The Internet of Everything 54
Artificial Intelligence(AI) 64
3D Printing 78
Robots and Drones 85
Blockchain 96
Augmented and Virtual Reality 105
Quantum Computing 114

Step 3 - Applying New Technologies to Create a Better and Smarter World 121

The Smart Home *and Why It Matters* 122
Smart Opportunities to Help People with Disabilities! 130
Smart Personal Transportation - Faster and Better 141
Smart Supply Chain Our Chain of Life 152
Smart Sports, eSports and Wearables 162
Smart Digital Communications Tools and Technologies .. 170

Step 4 - Applying Process for Customer Business & Career Success 179

Why It is ALWAYS About Your Customer 180

Opportunities & Success in Industrial-Age Companies 187

Why You ALWAYS Need a Plan 194

How to Generate an Endless Supply of Digital Ideas, Solutions and Prizes 201

Why Digital Strategies are Different and How to Profit from Them 210

Why Digital Business Models are Unique and How to Profit from Them 217

How to Create Value Fast Using MVP's and Pivots ... 233

How to Leverage Content, the Currency of the Digital World! 238

Why Data is the New Gold! - How to Manage, Value and Protect It 253

How to Buy Digital Technology, Be Happy and Never Get Burned! 265

How Two Technologists are Connecting Programmers Worldwide 269

How an Innovative Recycling Company is Changing the World 276

How a 185-Year-Old Railroad is Creating Digital Delight 281

How Grandma Became an Unlikely Digital Success Story 285

About the Author 292

Step 1

Developing the Foundation for Digital Success

This section, in combination with emerging technology and process information in the following sections, are the keys for you to enjoy digital business and career success.

In this section, you will learn:
- Why this a unique time in history to digitally transform the world
- 22 positive habits that drive digital success
- 7 negative habits that cause failure
- How to look at the world on a slant, discover and monetize hidden value at your fingertips
- How and why to 'fail', in order to succeed
- How to use automated tools to digitally 'listen to' torrents of raw data and easily identify well hidden 'needle in a haystack' opportunities and threats
- How and why 'digital play' is a key to success

For your convenience, up-to-date links to all companies and products discussed in this section plus new golden digital nuggets are listed on DigitalSuccessBook.com/links.

Rocking the World with Digital Transformation and Convergence

In this chapter, you will begin building your digital transformation foundation. You will learn how rapidly the world is changing and why digital convergence is the fuel for change. By the end of this chapter you have a high-level understanding of the building blocks that make the emerging technologies possible. Later, in Section 2 we will go into detail about the value of each technology and how it can benefit you.

Let us begin our amazing digital journey with a thought exercise. Which of the following products and technologies do you think are currently in production or in the testing phase?

- Self-driving cars and trucks
- Flying Taxis
- Autonomous planes and helicopters
- Machine that teach themselves. One that beat the best Go players in the world, which is exponentially harder than Big Blue beating Gary Kasparov in chess.
- Machines that think creatively
- 3D printers that print cars, planes and custom prosthetics better and at a fraction of the cost of existing manufacturing processes
- Artificial intelligence that partners with doctors and lawyers on research and decisions
- Dynamic product pricing online and in physical retail stores
- Computers that control the direction of the stock market and the investments of millions of people
- Personal robots that enable seniors to live in the comfort of their own home for life if they so choose, instead of nursing homes.
- Robots that do dangerous work in factories and warehouses so humans do not have to
- Robots and drones that save lives during disasters
- Inexpensive office robots that work safely side-by-side with accountants, purchasing agents, HR,

customer service and most any other white-collar office jobs.
- Nano (teensy-tiny) machines that build structures at the atomic level and can even work inside our bodies to find and fix what is wrong with us.
- Miniature blue-tooth enabled pills which we can safely ingest that can provide critical data to medical professionals in real-time
- Miniature devices we carry in our pocket, and would never leave home without, that are more powerful than the mainframes that landed people on the moon.
- Drones used for security, work, medicine and supply delivery during and after a crisis
- Computers over 1,000,000 times more powerful than today's supercomputers that will enable us to do the unimaginable

Incredibly, everything listed above exists today or will be here very soon! Throughout the book, we will discuss all of these and so much more but before we journey into why everything above is becoming possible, one more quick thought exercise.

Make believe it is 1995 (even if you were a toddler or not born yet) and we are sitting having coffee in a 'coffee shop or diner.' Those were the options for coffee consumption back then. The coffee industry had not yet been disrupted by Starbucks, Gregory Coffee and other amazing ways to consume 'my beverage of choice.' Also, Keurig machines in the home and office appeared years later.

During that 1995 discussion, I told you I had a dream last night that in a few short years our world would be reinvented and I went on to list some of the items on the list above, that the way we win at business, get entertained and improve our health would be radically changed and that most of the current (1995) industry leading companies would be displaced by 2018. In addition, many of the 'displacers' would be started in garages, cubicles, college dorms and basements. How coocoo and ridiculous would you have thought I was?

Flash forward to 2018:
- Newspapers: When was the last time you waited until tomorrow to get sports results, stock prices or the 'breaking news'?
- Photography: When was the last time you went to a 'camera store' to have your pictures 'developed'? When did you last see a Kodak or Polaroid commercial?
- Entertainment: When was the last time you popped a VHS tape into your video player? Probably not in the last 5 years except perhaps to watch family videos. When was the last time you were late returning a video to Blockbuster or Hollywood Video and had to fork over the 'late fee'?
- Disaster Recovery: If your business has critical data, when was the last time you backed the data up to tape? I hope you say a long time ago, otherwise please buy my Amazon Bestselling - 'The Ultimate Business Continuity Success Guide' (shameless plug).
- Communications: Perhaps, you remember calling loved ones or business associates sparingly back-in-the-day as the fees were a dollar a minute and much higher. When was the last time you thought about the cost of long distance communication fees?
- Floppy disks and AOL free DVD's: Remember those? Those relics of the past are now called 'coasters'.

For your organization to thrive and survive, understanding how to leverage emerging, exponential and empowering technology is critical. The need to accurately predict technology trends in the future is key to building exciting new digital products, services, platforms and ecosystems that will delight customers and enable you to win the digital war.

Would you rather languish in an old-school dying industrial company or excite customers by building better, faster and cheaper products and services? The Industrial Age is fading away. It was linear whereas digital is exponential in every way. In my experience the advantages of thinking and being digital make it an unfair fight. Digital wins every time!

Understanding your customer, having the right digital mindset and realizing the amazing technologies available to you are keys to the kingdom. This chapter will explain why hyper convergence is upon us, why it will continue to get better and why it will not stop in the future.

Later in the book we will visit innovators making the most of digital transformation. Some are disrupting their industries. They will share their secrets to success. These are not the usual mega-titans you are familiar with such as Musk and Bezos. I deeply respect them but you already know their stories. The people in my book are lesser known but doing as great things as the headliners. They are making our world a better place in their own unique ways.

As you read through this chapter begin thinking about these questions:
- What are your ultimate goals?
- How can new technologies help your business or career?
- What do your customers or employer want that they do not have and nobody can currently give them? Don't hold back! Digital technology can take you anywhere you want to go and I will show you how.

The Laws of Exponential Technology Growth in a Nutshell

The digital world is changing incredibly fast. We are now using products and services that were only possible in science fiction movies a few short years ago. It is the exponential advancement of technology that has driven this type of change and opportunity.

You can build and deliver amazing products and services whether you are a sole digital developer working in a spare room (been there / done that) and/or a corporate technologist working at a firm with thousands of people (been there / done that) or as a business officer or director (yeah, been there and done that as well).

Technology has been growing exponentially rather than in a linear fashion for many decades not just the past few years. Never make the mistake of thinking that it will stop or slow down. Technology will only get better, faster and cheaper.

I have always been keenly interested in the compounding effect of exponential growth. When I was in school I had a finance course that did not particularly excite me. One day the instructor talked about compound interest and the 'rule of 72'. I lit up like a smart bulb (more on those later). The professor was shocked at my enthusiasm. I totally got it. Technology is like compound interest, only better.

Prior to the recent digital convergence many of the automation and process improvements impacted businesses internally behind-the-scenes. I built a lot of internal solutions that benefited corporations and their shareholders. Many of them are in use today.

Somewhere around 2016 technology achieved an inflection point. The core building blocks of digital technology that I describe below had matured to a degree that they now empower us to create futuristic products and services.

Importantly and contrary to what was thought to be possible, as technology improved it got faster, better AND cheaper. It defied what marketing gurus thought was possible. Just as in programming Boolean logic, there is world-changing difference between the 'or' in the first scenario and the 'and' in the second scenario. It is world changing to have the 'and' to build with instead of the 'or'!

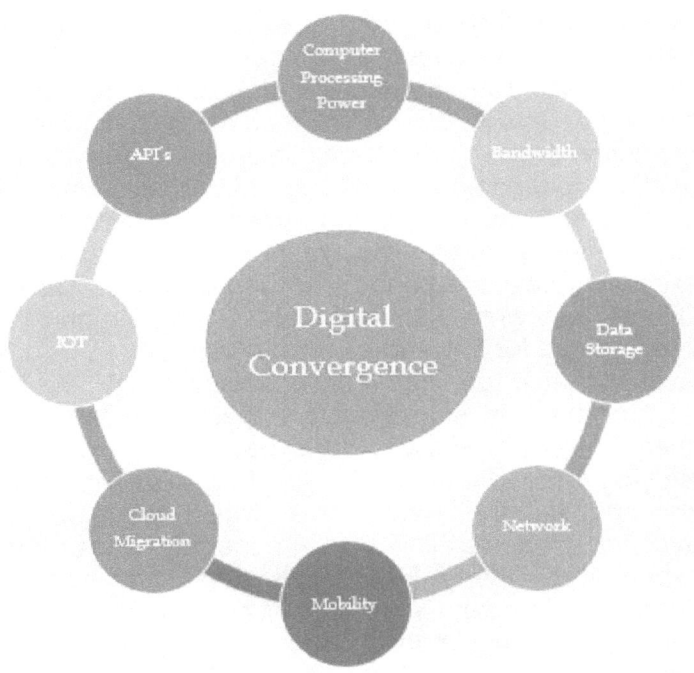

Convergence

Let us now discuss and demystify the exponential technology building blocks that are cornerstone to our new digital:

Computer processing power:
The brain of a computer is the microprocessor chip. As it improves, software leverages the additional power and more complex solutions can be developed. As of this writing the fastest supercomputer can process 200,000 trillion calculations per second. First hardware is developed and then software is written that takes advantage of the hardware improvements.

Processing power has improved exponentially over many decades. Computer chips are measured by speed and how many transistors they press on a silicon wafer.

As the size of the transistors rapidly became smaller and more efficient, more of them could be packed onto a wafer. The cost also became exponentially less.

No conversation of computer processing progress is complete without a short discussion of Moore's law:
On April 19, 1965 Electronics Magazine published a white paper by Gordon Moore, co-founder of Intel. He made a prediction about the semiconductor industry that has become legendary. Specifically, his Moore's law says the number of transistors on integrated circuits doubles approximately every two years. David House, an Intel executive, said the period was approximately "18 months". Overall, they were both right and this amazing prediction has been correct for 53 years and counting.
Recently the popular consensus is that the rate of advancement is slowing down to approximately doubling every 2.5 years. Some 'experts' have even said Moore's law is dying or dead. I totally disagree! I believe Moore's law will continue for the foreseeable future and around 2025 everything that we think of as exponential will become super-exponential. 'Marty Fox's' law says that when, not if, quantum computing makes its way from the laboratory to desktops and our mobile devices processing power, speed and the types of solutions we will enjoy will be 1,000,000 times better than what is possible now. We will journey to the world of quantum computing in Section 2 of the book and I will describe some applications now in production.

The evaporating cost of producing a microchip:
In 1958, a scientist at Texas Instruments developed the first-ever integrated circuit. It had two transistors (the more, the better) with a "gate process length" (the smaller, the better) of about ½ inch. This scientist would go on to win the Nobel Prize. Each transistor cost about $1, on average.
Now fast forward to 2012, Nvidia released a new graphical processor unit (GPU) with 7.1 billion transistors - the cost of a transistor: ~ $0.0000001. In that relatively short period of time chip technology experienced a 100 billion-fold improvement, right on schedule for Moore's Law.
In 2018, transistors the size of an atom are being created in research labs. As I said before, technology marches on, getting better and better!
In the chapter, 'How to Digitally Play, Prototype and Profit', I describe the Maker Movement and using inexpensive components such as miniature microprocessors. Arduino and Raspberry Pi are two examples of these super fun tiny,

powerful and very inexpensive ($5 – $40) computers. They are a direct result of the impact of Moore's Law. More (not Moore) on how they can be used later.

Bandwidth:
Data is made up of bits and bytes. A byte is the smallest piece of information people can understand. For example, each byte can represent a letter or number. There are eight bits to a byte. You can think of a bit as an on-off switch. Depending on the on-off patterns bytes are created. Bandwidth is the transfer rate of data.

In a relatively short period we have advanced from super slow 3600k baud 'data-thru-a-straw' 'to wide-pipe fire-hose data transfer rates measured in megabytes and gigabytes. Data transfer rates will make a giant increase to 5G in 2019 to 10 gigabytes per second. Data transfer cost has dropped to near zero. There are many reasons for faster and cheaper transfer rates including upgrades from copper wires to glass fiber optic cables.

The significance of faster and wider bandwidth is it enables the creation of customer delighting services such as Netflix, YouTube, and real-time gaming. Business applications will leverage this seemingly unlimited bandwidth with applications including immersive long distance meetings, virtual and augmented reality.

Data storage:
Exponential data storage capabilities went from single-sided floppies, to double-sided, to 3 1/2", to ten megabyte Bernoulli boxes (which were a beautiful thing) to terabytes and exabytes and big data cloud drives for pennies a megabyte. Storage used to be a concern but is now virtually free. Almost infinite storage capabilities enable YouTube to store millions of videos and Google to store millions of scanned books.

Internet
The global network gives us access to billions of connected people and in the near future trillions of devices. As the network grows and there are more people and devices

(nodes) added, it becomes exponentially more valuable. It has the same network effect as the old fax machines. When the first fax machine was sold, it was not very useful, as there was no one to send a fax to. As millions of fax machines became staples of most businesses the fax machine became a valuable business tool.

By the way, the fax is a great example of a formerly indispensable device on the wrong side of digital disruption. I am sure some of my readers are asking what a fax machine is.

The next step for the Internet will be going from Internet Protocol (IP) v4 to IPv6. This is very important. We will skip IPv5, which was an experimental streaming real-time streaming protocol. Each computer and IOT device we add to the Internet uses an IP address. You can think of an IP address as your unique street/city/state address or your telephone number. IP is the magical plumbing of the Internet.

Because of hyper-growth we will run out of the approximately 4.3 billion address capacity of IPv4 soon. IPv6 fixes that problem for the' foreseeable future' as it has a capacity of, hold onto your hat, 340 undecillion addresses. Ok, I never heard of an 'undecillion' until now and I cannot resist giving you a bit (pun) more info. Fun fact, this is the exact astronomical number of addresses that we will have available to us: 340,282,366,920,938,463,463,374,607,431,768,211,456.

That is far greater than all the grains of sand on earth and even greater than all the atoms on the face of earth.

Cloud Computing

The cloud levels the playing field for small companies. For a reasonable price, that gets lower every year, a small company or solo entrepreneur can have access to supercomputers and all the storage space they need. Services such as Amazon Web Services, IBM Cloud, Google Cloud, Microsoft Azure, Rackspace and Digital Ocean provide world class infrastructures. No longer are business required to purchase and maintain costly infrastructure hardware and hire people to maintain it. The cloud is generally more economical and secure than maintaining your own local data center.

These super cloud platforms also carry a lot of panache with corporations. When I am evaluating a 'software as a service solution', I always ask the vendor who they are hosting with and usually I get the right answer, which is one of the previously mentioned cloud services.

Importantly, cloud migration also means that our local devices such as smartphones, Chromebooks and tablets do not require expensive internal processing power to do incredible things such as real-time language translation, artificial intelligence and realistic simulations. The heavy-duty processing can be done on supercomputers in the cloud that are only a click away and the results are then displayed on the local device. If your company will be developing mobile apps you will want to consider where it makes the most sense to do data processing and storage.

Mobile:
The miniaturization of devices is changing our world for the better. Our smartphones have at least 11 sensors in them and more computing power than the supercomputers that only a few decades ago took up an entire floor of a building and put people on the moon.

Travel directions and the ability to order anything from a taxi to a pizza is in your pocket or pocketbook. When was the last time you fumbled with a crumpled-up 2-foot paper map? There are now millions of apps that make our lives easier and more productive. People who previously could not afford a computer can now afford to put a supercomputer in their pocket or pocketbook.

Your business depends on your mobile capabilities. Mobile apps have saved me many times when sending a critical mass notification to employees during a crisis event from a donut shop or library to thousands of employees. As our technology becomes exponentially more powerful our devices become smaller. They are already tiny enough to be woven into our clothing, imbedded in our skin and swallowed in blue-tooth enabled pills. We will discuss the benefits of these products and learn who is making them in the technology section of the book.

Think about how fast miniaturization has shrunk our computers. A few short years ago, we all had similar beige desktop computers that were anchored to our desks. Yes, every desktop had to be beige and it had to be tethered to a desk. When Osborne and Compaq released 'luggables' (35 pound 'portable' computers) it was both a breakthrough and a back-breaker. I remember going on vacation with a luggable and a suitcase full of 500-page technology text books - total weight over 50 pounds! The luggable still had to be plugged into a wall socket or it was just an expensive workout weight. It would have taken a car battery to power it, and yes that experiment crossed my mind briefly.

Now we can easily carry thousands of times more computing power using mobile devices weighing ounces, not pounds. I will not insult these devices by calling them mobile phones as they do so much more. In addition to a mobile device, you can bring along thousands of books, documents and magazines on an eReader. I am sure many of you are reading this book on your beloved Kindle.

Internet of Things (IOT):

IOT gets me extremely excited. It consists of billions of value-laden smart sensor enabled devices that allow us a deeper understanding of the world, whether it's for health, sports, entertainment, or business. It is a game changer for innovative organizations.

IOT opens up a world in which digital and physical elements connect and gather information in real-time to predict circumstances, prevent problems, and create immense opportunities and benefits for our society.

Internet connectivity and smart sensors are being integrated into industrial processes, transportation routes, workforce practices, buildings and every type of operational system. This is improving and revolutionizing the efficiency, productivity, and effectiveness of individuals, businesses and governments. The Internet of Things – or the Internet of Everything - is transforming the way we work and play in today's world.

Application Programming Interface (API's):

I strongly feel API's belong on this list of critical technologies. They are the connectors that allow us to mash and mesh systems and data. They allow machines to talk to other machines. They provide the 'sum is greater than the parts' building blocks we need to create simple and complex systems.

They enable you to connect applications to provide customer delight and create an infinite number of innovative solutions. I like to use the analogy of snapping together Lego blocks or Tinker toys to create amazing structures. By the way, if you have a child that loves building with Legos or other such blocks, it is an excellent sign of good things to come. I speak from experience as the proud dad of my favorite professional software developer, engineer, technologist and son. He was a Lego master builder.

As a bonus, in 'The Internet of Everything' chapter, later in the book, I am going to share some amazing new middle-ware tools that make using APIs as easy as inputting a few pieces of information and snapping systems and physical devices together in minutes. It is magic and no technical skills are required to do it. How cool is that!

I hope you are ready to Digitally Rock the World!

22 Habits for Digital Success

Technology is the great enabler. Unfortunately, many people that understand technology from a nuts-and-bolts perspective do not have winning digital habits. Some are sitting on goldmines of opportunity and others are the targets of disruptive threats but they just do not see 'through the fog'. Some do not know how to, some do not have the time to and others just want the world to stand still. They never execute a digital success plan. Instead they set themselves up for failure.

Developing the right mindset and habits is key to digital success. How you identify opportunities, threats and solve problems will position you to create customer delight and a successful career.

I live by these habits every day. They have been good to me for many years and enabled me to find success in business, career and my personal life. Allow me to share them with you.

The 22 Habits that Can Change Your Business and Career:

Be a bold leader, not a follower:
Lead by action. Create a culture that values innovation, creativity and thinking digitally. These attributes should be core to your team's organizational values. Motivate, incentivize and simply 'light up' your teams and peers to the incredible digital opportunities to transform your business. Doing so will empower you to grow a successful digital organization! You will be setting the stage to disrupt and to never get disrupted.

Be optimistic and maintain a positive outlook:
An optimistic and positive outlook are important ingredients for digital success. Too often people take a negative attitude and beat themselves when success is in their grasp. Optimists positively affect all those around them. Every conversation I engage in and every project I am part of I exude optimism. I am proud to say that I have often been called by co-workers, teammates and friends the most optimistic person they ever met. If you are always positive you will rarely have a bad day.

Whether things are going well or not, stay positive and project a positive aura. Digital transformation and disruption will bring challenges but that is great, we embrace challenges!

Imagination is more important than knowledge:
Here are two quotes from Albert Einstein:
"Knowledge is limited. Imagination encircles the world."
"Imagination is more important than knowledge".

My take on it is- if you can imagine it, you can do it. What you can do with exponential digital technologies is only limited by your imagination and you do not have to be a programmer or a tech genius to be digitally successful. 99.9% of customer delighting digital opportunities are waiting to be created and brought to market. Imagine it and create it.

Create greatness, not mediocrity:
Reach for greatness. Do not be conservative. Never be mediocre. If there is a disruptor that is entering your niche don't try to equal them, rather leap beyond disruptor's and squash disruptor's. I promise you anything is possible if you focus, engage your teams and leverage powerful emerging technologies and tools.

If you are an employee think like an intrapreneur. An intrapreneur means having an entrepreneurial attitude while working for a company. Your employer will love you for it and should reward you. Treat your daily job as a business that you must improve to generate success. Seek out ideas and tools that can differentiate your organizations and bring them to your management's attention.

Think 10x and monitor the edge of your sector for opportunities and threats:
Think BIG! Go on the offense. Disrupt, to prevent being disrupted. My biggest successes are often the ones that are my own little 'moonshots'. The ones that are so different they create a 'market of one'. The ones that some people said were 'a little crazy'. Those are the best. Imagine teaming with a laid off grandmother who had zero technology knowledge to build a social media site? I embraced that opportunity, we helped thousands of people, it became

famous and was featured in The Wall Street Journal, People Magazine and many other media outlets.

Brainstorm with colleagues on how your organization can disrupt and leap beyond the competition. All ideas should be respected and considered. If you are a manager you want people to feel comfortable and safe providing ideas.

Google walks-the-walk and talks-the-talk; they provide ample time for employees to work on projects outside their normal duties, on company time. The last time I looked, it was 20% of their time. Employees enjoy the perk. Google gets better employees to work with them and revenue generating products are derived from the program. I call that win-win-win. Wouldn't you want to work for a company that allows you to follow your passions? Be that company!

Embrace and create a culture of change:

Everything is on board for radical change. Forget the same-old-same-old. Some people will insist on doing things 'the way we have for the last 20 years'.

One of the biggest challenges in implementing a new technology or process is change. Change creates a multitude of feelings; for some it is apprehension and uncertainty.

It is important to create a culture that embraces change. The first step is to lead by example. Always seek a better way of doing things. Push new ideas, concepts and processes. Anticipate resistance, especially in industrial era culturally siloed organizations. To become a disruptor everyone in your organization must buy-in. Discard the naysayers. Do not let them bring you down. If they persist - it may be the highway for them. Change is a team sport. Put another way, there is too much dead wood in companies - especially in large ones where it is easier to blend in or hide.

Seek and embrace challenges:

John Kennedy said, 'We choose to go to the moon, not because it is easy, but because it is hard'. This book is focused on finding ideas and challenges and making you into a disruptor.

Innovation and digital disruption is centered on finding challenges and overcoming them. Become a problem solver.

If you have the right approach to challenges you will embrace them and covet them. Overcoming the right challenges can make you rich, famous and more importantly empower you to help people. I feature some of those brilliant innovators that make the world a better place later in the book.

Life can become boring if you do not get out of your comfort zone. Whether you are an owner or employee, making yourself uncomfortable allows you to break new ground and grow. For me, it is writing a new book after exhausting myself on the previous one, starting a new digital business, giving speeches and doing large scale webinars. Taking on seemingly insurmountable challenges is exhilarating and, at least for me, what life is all about.

Be resilient:

Organizations and their employees must be resilient. Resilience is the ability to face any type of challenge, even the most disastrous and disruptive and bend but never break. It is the ability to absorb whatever happens and spring back not only to the previous state but to learn from the challenge or disastrous event and become even stronger. Much of my previous book, The Ultimate Business Continuity Success Guide: How to Build Real World Resilience and Generate Profit Streams was centered on building resilience for organizations and their employees.

Have a great athlete's mentality:

Derek Jeter and Michael Jordan had it when they were playing and now use it to their advantage in business. They were the first to arrive at practice and the last to leave. They 'practiced with a purpose'. They prepared meticulously covering every detail. In my own small way, I have always approached everything I do that way. You cannot always be the most talented person but you can be the hardest-working.

Digital transformation is a never-ending journey. What you implement now will be old-school a decade or less from now. Always strive to improve. I think I had some of that attitude instilled from a lifetime of competing in sports where things did not always come easy to me. I was told by my friends that 'I was never satisfied'. I prefer to call it an athlete's mentality.

For example, a couple of years ago, I fell while playing basketball outdoors and hit my head on the concrete floor. It ended my long basketball career. Fortunately, I used my athlete's mentality and became a race walker. I practiced with a purpose, relied on data generated from my fitness tracker and experimented relentlessly. I 'failed fast and often' in the beginning, but eventually everything came together and I won the U.S. Track & Field National 10k Championship and came in second in the National 5k Championship. I then transferred my new-found knowledge to running local 5k's and 10k's and it worked out very well. But I still want to improve. I enjoy competing against myself and the clock.

Never give up the digital fight:
The stakes are too high to quit on innovation, digital transformation and disruption. As Mo Rocca says on Henry Ford's Innovation Nation, one of my favorite innovation TV shows - "NEVER GIVE UP!" You will make a better future for your customers, organization and yourself! I believe if you do not quit you cannot lose. Happily, Edison, Tesla and Colonel Sanders felt the same way.

Question everything!!
As you will learn throughout the book anything is possible. Digital transformation is about creating new products and services using exponential technology. It is also about relentlessly automating industrial age products and services. Every product or service can benefit by adding intelligence and becoming smart. I devote a section of the book to creating smart homes, cars, aircraft supply chain, smart devices to help people with disabilities, communications, entertainment, sports and more. Each chapter will help you develop ideas for products that are not yet smart but ripe for added intelligence through digital transformation.

Be different:
Never be afraid to be different. Think 'outside the box' and 'take the road less traveled'. Go far beyond the disruptors, instead of trying to copy them and compete with them. A 'follow the herd' mentality no longer works in the digital world. If you are trying to copy someone that has first-mover advantage in your niche you are making a mistake.

During the industrial age 'borrowing ideas' might have worked. We all know companies that copied the little creative people and spent a lot of money on marketing and advertising to crush them. In the digital world that no longer works. The Internet has leveled the playing field. It no longer takes a big budget to market and create awareness. It takes a great idea, product or service and the ability to execute.

Discover riches in the niches:

John Wanamaker (1838-1922), department-store titan, once said, "Half the money I spend on advertising is wasted; the trouble is, I don't know which half." Mr. Wanamaker would have loved the digital world, as there is no longer advertising waste.

The Internet gives you the ability to target niches. Kevin Kelly's book, 'The Long Tail' pioneered the subject of niche marketing and it still holds true today. 'The One to One Future' by Martha Rogers and Don Peppers was ahead of its time and a great book. I have written about and successfully used targeted niche marketing many times.

You need to define your customer persona, the characteristics of people that typically buy your product. It is then a matter of being a detective and finding those people. I am a student of hunting and prospecting and I provide lots of tips in the book for you to successfully locate people and organizations that can benefit from your product or service. In one instance, it allowed me to help build a two-person software company into a multi-million-dollar success.

Be agile and lean:

We will discuss many of the advantages of assuming an agile mindset and a few things to be wary of later in the book. Often small teams win, while big teams bog down. Cut meetings to a minimum. No meetings should be over an hour unless necessary. Reduce risk using hackathons and platforms we discuss later in the book to build better, faster and cheaper solutions.

Build with a sense of urgency

I see it way too often. An organization decides to implement a new digital transformation program. They are gung-ho and say all the right things, but move too slowly. Management interest and support begins to wane. There is not that sense of urgency that a fast-moving digital program requires. Unfortunately, the people that want to disrupt you are moving quickly and if can spell disaster if they get to market before you do.

Pivot when it makes sense:
Pivoting products or services has turned mediocre companies into digital giants. I will share specific cases later in the book.

Encourage experimentation and covet failure:
The importance of relentless experimentation and coveting failure are discussed in detail in upcoming chapters including, 'How to Discover Extraordinary Value in Experimentation and Smart Failure" and 'How to Create Value Fast, Using MVP's and Pivots'.

Have a 'childlike curiosity':
See the world as you did when you were a kid. There were no preconceived limits. You played and experimented, made up stories and had fun. Everything was so new and exciting. You figured out how to do things in creative ways. Well, that is what your role as a digital leader is all about.
Einstein said of curiosity, "The important thing is to not stop questioning. Curiosity has its own reason for existing."

Read-Read-Read and Listen-Listen-Listen:
It is critical to keep your eyes and ears attuned as the digital ecosystem moves fast. Throughout the book, I share my secrets on staying up-to-date on digital products, services, opportunities and threats. Survival and success means staying ahead of the competition. I will show you techniques and tools that will make a difference.

Don't reinvent the wheel:
Digital lends itself to re-use and easy customization of content and code at little or no cost. Bits and bytes have that wonderful quality. Throughout the book, we will discuss reusing bits and physical atoms.
I will also share the inspirational story of how TerraCycle, a sustainability leader, turns worm poop, cigarette butts, plastics and other garbage into valuable products and most

importantly, makes the world a cleaner and healthier place to live.

Play and Have Fun!

Again, I will rely on an Einstein quote to say it better than I can: "Creativity is Intelligence Having Fun"

Michael Jordan on playing basketball - "Just play. Have fun. Enjoy the game".

Marty Fox - "Digital Innovation, Transformation and Disruption (if you are doing the disrupting) is fun stuff. Playing and having fun leads to success in every facet of life."

I will share the importance of playing and having fun in the upcoming chapter, 'How to Digitally Play, Prototype and Profit'.

Love what you do:

The adage that if you love what you do, you will never work a day in your life, is so true. I love working with digital technology and applying it to transform businesses and lives. It may be a bit extreme, but every day I eat a bag lunch at my desk, except for the rare occasion I am attending a business lunch. I rarely take days off because I enjoy my 'work' more than anything else I would do. I get some flak for all of this, but I am comfortable 'in my skin' and have learned to accept it as who I am.

The best advice I can give you is this; if you are going to a job that you despise, if you are stressing out or depressed, if you have difficulty sleeping with the thought of facing another day of work it is not healthy. Find another job or career. Going to work for 40 years so you can start enjoying life at 65 is not a workable plan.

7 Habits for Digital Failure

I hope you were inspired by the preceding chapter that described my 22 habits for digital success. The 7 habits for business and career failure described in this chapter are equally as valuable. I hope understanding how toxic they can be enables you to avoid their pain.

Unfortunately, the story I describe below is all too real for many organizations that attempt to digitally transform. Over 80% of organizations report their digital transformation efforts either fell short of expectations or completely failed. Throughout the book, we will focus on fixing this so you are in the small percentage that creates extreme customer value.

Once upon a time I had the opportunity to listen to a rousing digital transformation speech by a forward-thinking high-level executive at a corporate off-site meeting. She is a leader of a multi-billion-dollar retail company primed to either disrupt their industry or be disrupted by outsiders.

I got excited listening to her speech which focused on delivering customer delight and simultaneously removing friction from their complex and too siloed global supply chain. She spoke of revolutionizing their core products and implementing an end-to-end digital transformation of their manufacturing and delivery processes. She spoke of their new corporate values centered around digital transformation and creating customer delight. She spoke of how everyone needed to 'think digital'. Her plan was a tall order but her passion for digital transformation was evident and that was a great start. Everyone clapped and cheered.

Unfortunately, the subsequent weeks and months were quiet and there was no action on the part of the troops to drive her vision to reality. Every passing day of inaction brought them closer to being disrupted and destroyed. Sadly, there was no sense of urgency and her vision bit the dust before it ever had a chance to flourish.

Here is what occurred in one of the many breakout subgroups she created. It parallels what happened in similar groups throughout the organization:

Her Digital Vision: 'We will deliver real-time key performance indicators, break down silos, create real-time automated data streams horizontally and vertically, improve data quality, eliminate needless redundant efforts and minimize manual mistakes'.

Old-School Reality: In breakout meetings, the troops decided that their current capabilities only enabled them to deliver metrics quarterly. Their plan was to collect data using 20+ spreadsheets, manually combine that data in one humongous spreadsheet and then do some spreadsheet analysis. There was no thought of creating and reporting real time metrics in an entirely different digital way.

There is a much faster and better digital way they could have approached the problem that would have provided real time metrics, minimized errors, reduced expense and increased revenue, which was exactly what the forward-thinking executive wanted to accomplish. The tools they needed were available but they never found them. They would have become the disruptors and their digital team would have been super heroes instead of being on the wrong side of disruption.

Unfortunately, for many reasons we will discuss throughout the book the scenario I described is typical of what occurs at too many companies in too many industries. The 80% digital transformation failure rate I cited earlier does not shock me.

Fortunately, it can be corrected with a simple mash-up of digital: thinking, vision, ideas, processes, emerging technologies, tools and importantly a realistic plan for execution - all of which I will share throughout this book.

7 habits of owners and employees that will lead to failure:
1. Assuming what always worked in the old days will work now and in the future. Digital is different. Big dogs are falling every day. Toys 'R Us announced they are closing all of their stores and 33,000 people are losing their jobs. Many Sears stores closed a few weeks before and Whole Foods was bought be Amazon. In an upcoming chapter on disruptor's, I list other well known cases of disruption. Be paranoid and be prepared.

2. Fear of rocking the boat: If you do not rock the boat, the boat will be rocked by a disruptor and you will be in the water without a life preserver.
3. Fear of new technology: Technology is your friend. Too many people think technology is going to displace them. You must partner with technology to succeed.
4. Fear of being different: Too many organization are copy-cats. It is a losing proposition.
5. Using 'hunch decisions' rather than 'data-driven decisions'. Good data rules! It is the truth.
6. Creating an organizational culture that only looks inward instead of both outward and inward: Being myopic makes you vulnerable to the disruptor in a distant sector salivating on easy prey. Be aware. Look outside your industry and within it. The company that can disrupt you may not be your current competitor. The disruptor can live in an industry far from yours. Technology can help you monitor millions of dynamic data points in your industry and beyond.
7. Destroying your employees spirit of innovation: If you have that rare person with an innovative spirit on your team do not beat the person down. Do not ask her or him to 'reign it in' because what she or he is doing is not a watered down 'repeatable process'. Make it a repeatable process. Do not say, 'this is not the Mary or Harry show'. Let your people express themselves, be different and shine. If you do not keep them happy your competition will find them or they might go out and find your competition.

Discover Extraordinary Value in Experimentation and 'Smart Failure'

"I have not failed. I've just found 10,000 ways that won't work"
Thomas Edison

"Failure is never truly failure unless you get depressed, crawl under the covers and give up"
Marty Fox

Experimentation and smart failure by management and employees must be encouraged, expected and coveted on the road to digital success. Unfortunately, in my experience this is the opposite expectation in most Industrial age companies. People with good ideas are afraid to try new things because their ideas might not work the first time. They worry they will be failures in the eyes of their boss and co-workers and perhaps lose their job. Some managers forget the 99% of great things achieved during the year and ding you on the time you went beyond the same-old process and tried something new.

We must get over the 'failure thing' if we want to truly be innovative and digital successes

Successful digitally transformed businesses instill a culture that it is okay to experiment, fail fast, adjust and try again (iterate). You can learn a lot from 'failure'. Sometimes we call them 'lessons learned'. I speak about this later in the book in the chapter, 'How to Create Value Fast Using MVP's and Pivots'. Often this iterative process can be done quickly and on a small budget.

I covet failure. When I am interviewing people for a job I ask them about recent projects they participated in that did not work and why. If I am seeking someone for a creative, innovative, digital transformation type position and they tell me they have never failed, that is not the person I am seeking. The person I am seeking is someone that is imaginative, thinks outside the box and enjoys building new and better products and services. Failure is part of the game.

I created a short list of people that failed quite often. I hope it inspires you to be bold. I hope someday you and I are fortunate enough to join this list of 'failures':

- Thomas Edison - failed over 1,000 times in an effort to find the best filament when he was inventing the electric light bulb. He was a relentless experimenter. Every 'failure' and adjustment got him a step closer to success. Each step on his journey may have been minuscule but when he discovered that 'needle in a haystack' he changed the world with clean light. When a reporter asked him, "How did it feel to fail 1,000 times?" Edison said, "I didn't fail 1,000 times. The light bulb was an invention with 1,000 steps." Think like Edison and you will do great things.
- Mike Trout - is a baseball player who fails almost 7 out of 10 times when he tries to get a base hit. He makes millions of dollars for the 3 successes (.300) and is likely carving a path to the Baseball Hall of Fame based on intelligent failure.
- Giancarlo Stanton - is a baseball player who often strikes out over 100 times in a season - but when he does not strike out watch out because a 500-foot home run may be heading your way!
- Michael Jordan, Jerry West, LeBron James, James Harden and every guard that has played in the NBA since day one - all failures! Every one of them failed on most of their shots. None of them hit 50% of their shots over a season or a career. I just wish I had 1% of their game.
    - I love this quote from Michael Jordan:
    - "I've missed more than 9000 shots in my career. I've lost almost 300 games. 26 times, I've been trusted to take the game winning shot and missed. I've failed over and over and over again in my life. **And that is why I succeed.**"

The greatest sales person in the world can 'smile and dial' and cold-call one hundred prospects, get ninety-nine no's (and f!*k y!# then click) and perhaps get one yes. If that one yes leads to the sale of a big-ticket product or service it can make the salesperson wealthy. For example, imagine you are a digital entrepreneur pitching VC's and M&A's. One yes from current VC Marc Andreessen, the co-creator of Mosaic which was the first popular web browser, and you are gold!

Innovation, relentless experimentation and learning from 'failure' did not start yesterday or even 50 years ago. You can learn a lot from the past. Throughout the book I sprinkle 'Learning from the Past' profiles. Please do not skip them. One of my favorite 'failure' stories started over 100 years ago. Enjoy...

Learning from the Past: Emanuel Haldeman-Julius and his secret to selling 500,000,000 books and over 6,000 titles
As you read this story please map it to your industry and product. The process of successful experimentation is important to you, not necessarily his product.

Emanuel Haldeman-Julius was born July 30, 1889 and died July 31, 1951. He was a self-made man. As a child, he was an avid reader. His love for books, similar to mine, steered him to become a writer, similar to me. Eventually he purchased a newspaper and became a publisher. Next, he created a series called the Little Blue Books which would define his career.

The series which began in the early 1900's was very successful for half a century. Remarkably, his process for attaining success maps well to attaining success in digital world, which is why I profiled him for your benefit.

Here are three key reasons for his success:
1 - Emanuel took the time to understand what the public wanted to read. He did not publish what only he wanted. He used feedback and data to make decisions. I can attest that works. Out of the thirteen books I have written and published my primary concern is always understanding what readers want and need, before writing that book. I rely on data for my critical decisions, not hunches.

2 - Emanuel differentiated Little Blue Books by attractively pricing them at 5 cents a copy. Priced for inflation they would be comparable to some eBooks in our digital age. As you will learn later in the book digital

models allow us to sell products at very low or very high price points. We can even give digital products away for free to disrupt markets and become wealthy.

3 - All 500,000,000 of Emanuel's books had the same cover design, the same price and were the same size. The only variable was the title. It was his treatment of titles that demonstrates how lucrative it can be to embrace experimentation and covet failure. I can relate because 6 of my previous 12 books had the same, to be honest awfully designed covers (I may be the worst graphics person ever), but the titles were on target and honest. They all did very well. I do like the design of the book you are reading.

Emanuel meticulously tabulated data for every title in the series. As the number of titles surpassed 1,000 he had a wealth of 'big data', for his era. He was clever and leveraged the data in many ways. He was way ahead of his time.

He set a baseline that each title had to achieve 10,000 sales annually or it would be sent to 'The Hospital' to be fixed. He had a hypothesis, garnered from examining the data, that a key differentiator between a successful book and a poorly selling book was the title.

Each book sent to 'The Hospital' would be examined by three editorial assistants. From the data, they could isolate key words and rules that trended in successful books. They then brainstormed and came up with the best title based on the insight and used that title for a trial. Because they carefully recorded sales on each title, they could accurately track what worked and what did not. If a changed title did not work it would go back to 'The Hospital' for another fix.

Here are some examples of books that were admitted to The Hospital:
- Privateersman sold 7,500 copies in 1925 and 8,000 in 1926. After a title change to The Battles of a Seaman in 1927, it sold 10,000 copies.

- Fleece of Gold sold 6,000 copies in 1925. After a title change to The Quest for a Blond Mistress in 1926, it sold 50,000 copies.
- The Mystery of the Iron Mask sold 11,000 annually before a title change to The Mystery of the Man in the Iron Mask that sold 30,000 copies annually.
- The Nobody Who Apes Mobility went from near zero sales to 10,000 annually after changing the title to The Show-Off.

- Many of his other books increased in sales 400%, 500% and higher annually after having a title fix in 'The Hospital'. Some had to be changed up to six times to find the best title.

Emanuel also wrote a detailed biography of his life, which you may find interesting and can apply to your digital products.

Personally, as an author and publisher, the titles of my books are what I stress over the most. As Emanuel proved the title can make or break a book. When you pour months and years into writing a book and so much depends on the title it pays to craft the best one possible. My process is to develop 40+ titles using powerful key words. I then get feedback from potential readers on what would inspire them to pause and perhaps read the description and enjoy the first few chapters for free. I review all the feedback data and publish it. If the title does not work I send it to 'The Marty Digital Book Hospital'.

If you are interested in learning more about Emanuel I suggest you try - http://www.haldeman-julius.org/

Whatever product you sell or service you offer, you should experiment and embrace 'smart failure'.

Do You Know What Business You Are 'REALLY' In?

If I asked you, *'What business are you in?'* What would be your answer?

Think about it for a few minutes and write it down. It does not have to be long but now is a good time to do it. We will revisit it toward the end of this chapter.

The story below is based on an entrepreneur's experience in the food industry years ago, but even if you are in a different industry, it is relevant to both you and me as we develop our digital mindset and outlook on the world. This true story demonstrates how a slight shift in thinking can present vast, formerly unrealized, opportunities.

As you will see, sometimes you must peel the onion a bit to get deeper insight into massive opportunities and threats. In the digital or industrial world understanding your true value is critical to success. Why leave money on the table? Whether you own a company or are an employee this story can open your eyes to the transforming value that might be at your fingertips.

Now, let's go to the movies. Grab a small bag of popcorn and "Action!"

The Founder, a movie starring Michael Keaton, is about Ray Kroc and his significant influence on making McDonald's a leading fast-food chain. I enjoyed the movie very much. It is an interesting business case-study on how a small innovative hamburger stand in California became a world-wide enterprise. As you know, in the digital world one or two people with a great idea can also change the world. Later in the book we will meet some of them.

Ray Kroc was a hard-driving salesperson. The movie opens in the 1950's with him selling multi-mixers with 'mixed success'. Multi-mixers were an example of digital transformation of that period.

Somehow, Ray gets an unusually large order from a small fast food restaurant located in San Bernardino, California run by Dick and Mac McDonald. He phones them from the first phone booth he comes across (no he didn't have a cell phone, mobile was many decades in the future) to let them know they somehow mistakenly ordered several multi-mixers to make their milkshakes. Only it was not a mistake. The brothers proved to be disruptors of their era.

Ray is intrigued. He packs his car and drives (no hyperloop or flying taxis) from Nevada to personally deliver the multi-mixers to the brothers. When he arrives, he is mesmerized by their 'lean' fast food business model (Eric Reis of Lean Startup fame was not even born yet). McDonald's was delighting customers.

To make a long interesting story short, Ray suggests and the brothers agree to expand their business model by selling franchises.

Ray is passionate about the business and becomes adept at selling McDonald's franchises to people in all walks of life. Unfortunately, his contract with the brothers earns him a meager few cents on each hamburger sold and not much more. It seems somehow the more he sells the less he earns. At a point of desperation, he meets a finance maven named Harry J. Sonneborn. Ray and Harry hit it off and Harry winds up reviewing McDonald's cash flow and balance sheets.

In my favorite part of the movie during Ray's lowest point Harry asks Ray the seemingly silly question, "Do you know what business you are in?" Ray is dumbfounded and replies, "The Hamburger Business, of course."

Harry pauses and delivers the gold (I still get digital success chills just thinking about it), "No Ray, you do not seem to realize what business you are really in. You are not in the burger business; you are in the real estate business." This 'out of left field' broad thinking visionary insight backed by a big data' paper ledger and a brief explanation of how to turn his business into a goldmine instantly turned the light on for Ray (the scene is on YouTube).

Location is everything. Franchisees needed great real estate locations and real estate is in limited supply. Ray started leasing real estate and sub-leased it to franchisees at a healthy markup. Later he started buying real estate and leased it to franchisees at an even healthier markup. Whether a franchisee made

money or not, this new revenue stream flowed to Ray. It was brilliant, Ray made a
fortune and this type of thinking can benefit and transform you and your business.

Perhaps, now is a good time to revisit how you answered the question I asked you at the top of the chapter, *'What business are you in?'*. Over the next few days give it a lot of thought.

How would the big digital winners answer that question?
Are you thinking broadly enough about your true value to your customer and possibly customers in adjacent and distant markets?

Are you currently making use of all your assets? Do you have unused assets or byproducts you produce that can delight your current customers or future customers you have not yet identified? They do not have to be in your industry.
Digital companies that become disruptors often think in broad terms far beyond the same-old ways of doing things. That allows them to create unlimited organizational greatness and a great deal of customer satisfaction.

    For example, Lyft and Uber do not pigeon-hole themselves as taxi companies. If they did guess what they would be? Taxi companies! Rather, they define themselves as 'people moving companies', which opens up many options.

    Amazon is a great example of broad-based value thinking. They started by selling books and branched into many other products, but that is not their most lucrative revenue stream. Amazon was born as a data-driven technology business. To be the best they needed to build the best technology infrastructure consisting of quality data centers, servers and software. They learned how to scale to a massive size using thousands of servers. They learned about security and redundancy and guess what happed next...
    Using the scalable infrastructure, they run Amazon.com on, they created Amazon Web Services

(AWS), which has become their biggest revenue generator. On the AWS cloud platform, they host some of the largest organizations in the world. They also host many small and medium size businesses. As of 2018, they host over 1 million organizations and generate 12 billion dollars in annual sales.

I have always enjoyed this type of 'what business am I really in' visionary thinking. It has enabled me to build successful digital products and services for many years. It has paid extraordinary dividends in customer value and in process improvement leading to increased revenue and cost reductions. Wide-vision thinking provides me with so many ideas that I could never act on them all.

Sometimes I look at what other companies are doing and the byproducts they create and are not monetizing and that opens another opportunity. A great example of this type of innovative thinking is the story of TerraCycle later in the book. They built an amazing business and helped the world by recycling other people's garbage problems.

I apply this 'slant detective methodology' to many areas of business and life. Digital transformation and disruption are about seeing the possibilities, being innovative, thinking out-of-the-box and being creative.

For example, I am a digital technologist but I am also a Director of Business Continuity and Disaster Recover (BCDR). BCDR is a mixture of people, process and technology to insure companies can survive any type of disruptive event. Historically the profession was defined as simply to prepare, respond and recover from a disruptive event. Hmmm, that role might not fill my entire day. It also kind of labels me as an expense. But there is a different way of thinking about it which I detailed in my 2017 book, 'The Ultimate Business Continuity Success Guide'. It is a fresh visionary way of positioning the profession as a revenue generator and expense reducer. Apparently, other people liked my new value-based thinking as the book was very well received and quickly became a bestseller.

Instead of dryly describing my BC role and value as simply creating plans, writing business impact analysis, maintenance and testing I get much more traction from

the action statements listed below, which people can relate to (notice the parallel to digital transformation), find value and benefit from:
- The ability to respond to any type of disruptive event in an adaptive and flexible manner. I promote a culture that enables us to solve whatever is thrown at us. This insures we all keep our jobs.
- The ability to gain strength from any disruptive event we encounter and spring-back stronger than our previous state. Not the same as before – stronger and more resilient than before.
- The ability to clearly understand customers and management's pain and to relieve that pain.
- The ability to implement a high degree of situational awareness through tools and processes. This is critical to being proactive in risk identification, prevention and mitigation. It includes the ability to predict the probability of certain disruptive events before they become a reality.
- The ability to understand our organization from end-to-end, from the tip of multiple supply chain tiers to the customer and every step in between. This includes processes AND data flow. This will enable us to identify new opportunities for customer delight, revenue creation and cost reduction.
- The ability to deliver quantitative AND qualitative insights to management. This can lead to many surprising customer facing and internal opportunities.
- The ability to understand inter-relationships between all our company's assets. This can expose many previously well-hidden opportunities and threats.
- The ability to maintain a 24x7x365 world-class digital infrastructure where required. This is an expectation in many industries in a globally connected world. This can be a differentiator / money-maker in many critical time sensitive industries.
- The ability to clearly understand all threats, risks, vulnerabilities and opportunities in real-time and have the capacity to act on them. This enables us to serve our customers when other companies may be in crisis mode.

As a Digital Ninja, analyze what you and your business really do. You may find value that can benefit people in related and unrelated industries.

How to Disrupt the Disruptors

Digital disruption can either be a goldmine or devastating to an organization and careers. It depends which side of the fence you are on. In my day job, I get paid to prevent threats from becoming disruptions. It is a tough gig because out of the thousands of potential man-made and natural disruptions and disruptors, if one seriously impacts us is can be catastrophic.

Disruptions can cause companies to quickly go out of business and everyone can lose their job - including me. I tell my management if I do not prepare properly for disruptions and we are severely impacted they should fire me. Fortunately, I have never been fired.

The big digital disruptors:

I am sure Blockbuster, Borders Books, Barnes & Noble, Sears and the taxi industry would agree that disruptors such as Netflix and Amazon, Lyft and Uber hada devastating impact on their business models and employees. If you knew anyone workingfor any of those formerly profitable companies, hopefully they found a job elsewhere.

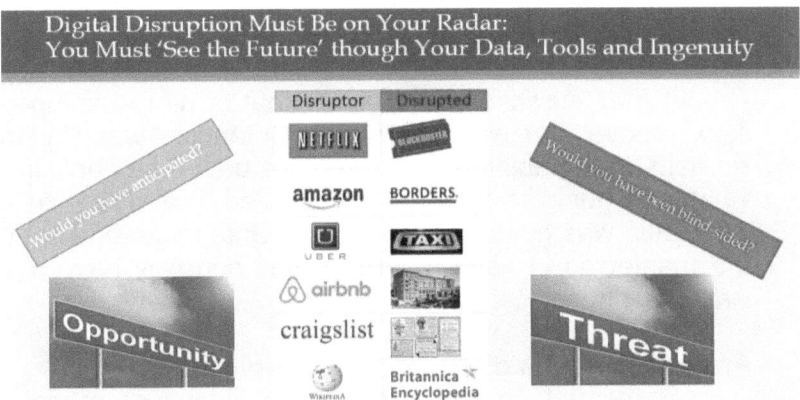

The interesting thing is that each of the winners (disruptors) listed above were built without the ownership of physical product. The barrier to entry was low because they used bits and bytes to develop unique value in the marketplace. They used the power of the network to reach prospects globally and virally, converting them into customers. Many of them used powerful new digital business models, such as free and freemium, which we will discuss later in the book. You can successfully use cool new digital models when dealing with the unique properties of digital bits and bytes, rather than atom based physical products. Customers flock to free and freemium. It is tough to compete with free.

Amazon makes buying a vast array of products frictionless with Kindle One Click and Amazon Dash buttons. In fact, per my recent Amazon bills, it is too frictionless and too easy! Do you remember traveling to the mall, fighting the crowds and finding that book you wanted just went out of stock 15 minutes before you got there? Gone is that friction. Already Amazon does same-day deliveries in many cities and soon they will be delivering using drones and their own delivery vehicles.

LinkedIn has been very successful creating a platform that connects business professionals. Initially, it disrupted the recruiting profession but recruiters proved resilient and turned lemons into lemonade. The savvy ones now leverage LinkedIn to source qualified candidates. LinkedIn even has a special plan for recruiters. It is a win-win situation.

It may surprise you to know that of all the companies listed above that were disrupted, Britannica was the most resilient, in my opinion, and fared the best. They anticipated what was going to happen and reacted well. They realized Wikipedia was gaining in popularity. Britannica may not own the market as in years past but they certainly have a very viable business.

Anyone can be a disruptor or disrupted:
To disrupt it takes a great idea, business model and the will to execute. Lean, agile, innovative people can quickly disrupt industries. It does not take a great deal of employees or much money. Even if some startup capital is required, simply go to Kickstarter. It does not take a lot of advertising. You can use social media, bloggers and targeted press

releases to reach a global audience. I provide an example of a press release 'gone wild' in this chapter.

Here are more examples of companies that have disrupted entire industries. Some are already household names. Others are up-and-coming companies. The industries they are disrupting may surprise you. Is your industry on the list?

Salesforce – disrupting the sales profession with their online platform and ecosystem. As an indication of their success they are buying prime office towers in New York, San Francisco and other big cities
Airbnb – disrupting travel accommodations. They own no hotels or real estate but are valued more than many major hotel chains
Coursera, uDemy – disrupting education
Square – disrupting electronic payments
Snapchat and Instagram – disrupted photography
Oscar – disrupting health insurance, health care
Apple – disrupting many industries including portable music players and the record industry
Tesla – disrupting car industry, battery industry, space exploration, travel (Hyperloop)
Waze – disrupting travel
Synack – disrupting cyber security
LinkedIn – disrupting the ways business professionals interact and the recruiting profession
Lyft and Uber – disrupting moving people from place to place. Much more than a taxi service
iTunes and Spotify – disrupting the music industry
23andme – disrupting genetic testing, ancestry
Craigslist – disrupting newspapers – those little classified ads were the biggest revenue stream for newspapers back-in-the-day, but not anymore

What should you be thinking about so you do not get a nasty surprise:
- Do you have a way of identifying disruptors in your industry or tangent industries?
- Does your company value and practice innovation?
- Could you possibly get blindsided today, next week, next month or next year by a couple of visionaries working in a garage?

Innovative startups can quickly roll out digital or physical products and grab your market share. The disruptive nature of technology will continue to become more powerful and available in the coming decades as the cost for the three most critical technology components: processing power, data storage and bandwidth continues trending to free or nearly free.

It is important for your company to have a pulse into what is occurring in your industry and beyond.

I use many digital tools and simple techniques to understand trends, opportunities and threats. A few of the ways this insight has benefited me are listed below:

- Teamed with a talented developer to market and sell one of the first live-video sharing applications. The first-mover advantage was key to our success.
- Developed one of the first Internet eCommerce engines. There was a significant need from web designers that wanted a simple to implement back-end eCommerce solution.
- Teamed with two entrepreneurs (husband and wife) from Europe that had developed a program called MindMan. I became their sole presence in North America after they had a bad experience with someone else. Together we made MindMan (now called Mindjet.com) a raging success story due to the quality of the product and creatively using digital intelligence and guerilla marketing.
- Built a popular social web platform that was acquired by larger company to achieve instant segment leadership. I describe the meteoric rise of GrandmaBetty.com later in the book.
- Discovered wants and needs using digital monitoring and wrote twelve popular books that satisfied those needs.
- Discovered trending subjects and popular search terms. Bought dot com (.com) domain names that mapped to those terms and built micro-sites to satisfy searchers.

Below are some techniques and tools I use on a regular basis day that enable me to know the past, understand the present in real-time and predict future trends and events.

Also, later in the book I will share how these and other methods help me generate an endless supply of digital ideas and revenue opportunities.

- Read and listen as much as you can. Whether it is books, social media, industry magazines, blogs, podcasts, videos, SlideShares or webinars, spend your time wisely and look for interesting opportunities and new products. Listen to customers, employees and friends. Information that was costly and difficult to find a few years ago, is now at your fingertips and plentiful, if you know where to look. I am a voracious reader and researcher.
- Attend or facilitate digital and physical meetups, conferences and conventions. I enjoy the large events and the tightly focused ones. I attend as many events as time permits and speak at some. These events offer great opportunities to network with like-minded professionals, learn about the latest happenings and speak with vendors - some of whom might be ready to disrupt. You can also learn of needs not being fulfilled in the niche which can be revenue opportunities for you.
- Use government sites and resources. Many, including Oak Ridge National Laboratory (https://www.ornl.gov/), NASA (nasa.gov) and DigitalGov.gov have outstanding technical information. They provide clues to what will trend next and in some cases, they may list government created products that can be licensed. If you are a citizen you are paying for this information, so make use of it.

Digital Listening Tools are Key:

There is too much new and changing information to monitor it manually. You must put yourself in position to discover the 'needles in a haystack' before the competition.

Listening tools can make the difference between being disrupted or being the disruptor. I suggest you consider using automated alerts, situational awareness, business and marketing intelligence tools. These automated tools can keep you in front of what is happening in our rapidly changing digital world.

My favorites use cutting edge technology including artificial intelligence and predictive analytics to make sense of torrents of public and/or proprietary data in real-time. Rather than simply passing you raw data, great automated tools add value to the data with filters and algorithms so the results match your business needs. They then push these insights to you, rather than you having to log on and pull the information.

When I started using these magical types of tools, I became so excited I passed on some of the insights to my friend who is a supply-chain professional. She told me it took her at least half a day of manual effort to discover what the tools produced in seconds. She would never have stumbled across the information that was automatically fed to her by the system.

Some savvy investors use intelligence tools in very creative ways. For example, a tool that provides weather information meant for crisis management may be useful to predict weather patterns. Perhaps, it can predict an upcoming freeze that will affect corn or wheat crops and subsequent prices. I use these tools to learn, in near real-time, of crisis events and to predict the likelihood of future events by researching their database of historical events.

In some time-sensitive cases, the intelligence you receive seconds or minutes before the competition can be monetized. Never has this been possible on such a grand scale. Intelligence tools also empower you to peer into the future. That can produce a competitive advantage for smart and creative companies that learn how to customize and creatively use the information.

Think about it, in a crisis such as a tornado wouldn't even a few additional seconds or minutes warning be potentially life-saving for your employees? During a recent tsunami ocean sensors coupled with notification systems saved many lives. I certainly would value that extra time to get my employees to higher ground and safety. It can be life-saving priceless information.

The information you act on must be accurate: Inaccurate information whether by mistake or 'fake news' can be worse than no information. My advice is unless it is from a credible source, you should dig deeper to insure the information is accurate.

There are many digital tools that can tip you off to threats and opportunities. A few that can provide value are:

Google Alerts (https://www.google.com/alerts) is a free resource that enables you to discover valuable ideas, news, contacts and trends. It searches millions of online resources on an ongoing basis and pushes precise information, that meets your criteria, to you in real-time or batched (sent in groups during at a specified time). You can set up multiple alerts on different keywords and text strings.

Hootsuite (hootsuite.com) enables you to monitor social platforms including LinkedIn, Facebook, Twitter, Instagram and Pinterest. It can be used in creative ways such as to generate ideas, monitor competitors, detect potential disruptors, collect information, automatically post content, analyze what people are saying about your company, discover career opportunities and anything else you can imagine.

Sendible (sendible.com) pro-actively monitors countless conversations on Facebook, Twitter, Instagram, YouTube and other platforms. This empowers you to discover what people are saying about your brand, company and competitors and allows you to respond timely and appropriately.

Google Trends (https://trends.google.com/trends/)is a powerful free trend discovery tool that enables you to learn what people are searching for, what is hot and what is not. It has valuable trending charts that enable you to anticipate what and when products will trend in the future. You can also use that information to discover disruptor opportunities.

NC4 (NC4.com) provides you with real time intelligence on man-made and weather threats. It is highly customizable by the type and severity of alerts. You can map threats to your assets. Although, it is focused on safety and security the alerts can be beneficial in many areas. For example, they can help you decide where to locate an office or store and where not to.

Early Alert (https://www.earlyalert.com/) is an excellent weather alert tool that is especially strong providing comprehensive management quality weather special reports during large events such as hurricanes.

RiskPulse (https://riskpulse.com/) has deep knowledge of weather and enables you to understand risks before you ship products. Threats can be mapped to stationary locations and mobile vehicles. Their weather information including forecasted freezes and major events can also be used to identify opportunities.

Twitter (https://twitter.com/search-home) is a gold mine of instant information but has historically been difficult to search. Their current search feature is a vast improvement over previous version. Twitter is a great tool to discover what is trending and the experts you want to follow.

If you are interested in learning more about using Intelligence tools in new creative ways I suggest you subscribe to my Digital Success Newsletter - DigitalSuccessBook.com/newsletterbonus

How to Digitally Play, Prototype and Profit

Everyone, including executives, entrepreneurs, employees and students in every grade should make time to play with cool new emerging technology. There is no substitute to hands-on experience.

Whether it is installing new smart home devices as I did below, creating interesting and unique objects with a 3D printer or experimenting with augmented reality apps, play and you will have fun and realize multiple business, career and personal streams of value.

This chapter is a perfect precursor to the upcoming chapter, 'The Internet of Everything', later in the book. IOT is already changing the world and we are just scratching the surface of what is possible.

The Do It Yourself (DIY) Maker Movement is here.
Building is fun. Some of my happiest memories are of digital projects and prototypes my son and I built beginning when he was 8 years old. We were makers before there was an official Maker Movement. I never had to push him, as he took to it like a duck to water. He loved building with Lego's and digital discovery was a similar creative endeavor. In elementary school, he was using early sensor type devices, simple timers and software to develop winning projects such as:

- Mail is Here: a prototype with a light sensor placed in a mailbox that was activated when the postman opened the box. It instantly alerted us that the mail had arrived. The benefit of this project was it saved time. We did not have to go out to the mailbox until the mail arrived.
- Bus is Near: a physical/digital prototype that used a small toy yellow school bus with a GPS sensor and tracker to indicate the location of the bus. The benefit of this project was children would not have to spend unnecessary time waiting in the cold, snow or rain until the bus was nearby. Why needlessly get wet or catch a cold if the bus was delayed.
- Electric Dog Feeder: simple timers were set on a dog feeding contraption built out of plastic junk parts. This

project went into 'Fox Family' production and freed us to go out without worrying about our dog being hungry.

My son went on to a very happy life as a leading software engineer. I attribute at least some part of his success to his early problem solving skill development and creating digital solutions in a fun environment. Oh, by the way the three projects I described above were ultimately built by innovators and made customers quite happy.

When you build something, even on a very small scale, you quickly understand the technology and the possibilities of where it can take you. In a couple of hours of fun and learning you can integrate smart low-cost physical objects, such as smart light bulbs, thermostats, cameras, and billions of other devices with free cloud platforms.

For example, you can save money on your electric bill by only turning on lights when you need them. You can install smart thermostats to save energy and make your home more comfortable. You can control objects from anywhere in the world. The night before writing this chapter I had dinner at a restaurant and I turned on the smart lights in my living room at dusk with the click of an app.

Beyond the immediate personal benefits, you will understand the power of integrating devices to improve offices, warehouses and factories. You will begin creating ideas on how radical new off-the-shelf technology can come into play. You may get an idea to disrupt your sector, an adjacent sector or a new unrelated sector. For example, below I share how the smart light project in my living room turned into some major career wins and will lead to many more.

Even if you are a C Level executive or founder, playing will make you more comfortable and confident as you lead your organization to digital greatness. I promise you do not have to be a 'hard core techie' to build solutions that can help you at home and can then be extrapolated to business value.

The global Maker Movement has been embraced by people of all ages. Many new products and services will be born from the efforts of 'regular people' with a passion for technology, design and lots of imagination. Some of these people you may want to hire, either now or in the future. You can find diamonds in the rough by getting involved with the Maker Movement personally or through your organization.

The Maker Movement started in approximately 2005, but the maker's innovative spirit goes back pre-digital for eons. Ben Franklin, Thomas Edison, Nikola Tesla, Dean Kamen, Paul Allen, and Steve Wozniak were all Makers. In some small way, my son and I were and are Makers. Chuck Pell, the terrific host of my favorite tech / future show XPloration 2050 describes himself as an artist, Maker, entrepreneur and futurist. How cool! I really hope you try being a Maker. All it takes is an hour or two to begin. There is probably a Maker Faire in your neighborhood or one that is sponsored by a local high school or college. It showcases innovators and entrepreneur's creations. You should attend one for ideas and possibly to find your next employee or partner.

If there is no Makers program in your city that can be an opportunity for you or your organization to start one. Participating will help you and your community. If you have children, you would do well to encourage them to attend to see what it is all about. As the proud father of a leading technologist and software engineer I promise you getting children involved with tech at an early age can be a very good thing for them.

Modern Makers use tools including tiny computers such as Raspberry Pi and Arduino that start at $5, inexpensive sensors, robotics and software to build anything. If they can imagine it, they can build it.

Smart Home Fun leads to Work Benefits in Less Than One Week

While writing this book, I was eager to begin building a smart home using off-the-shelf Internet of Thing devices, platforms and software.

The unlikely starting place for my home digital transformation was Walmart. Don't laugh. I know, it is not as hi-tech as the MIT Media Lab but smart technology is now mainstream and you just never know when inspiration will occur.

I was wandering the aisles and of course found my way to the electronics section. A new showcase caught my attention. It was filled with smart light bulbs, electronic sockets, thermostats and smoke detectors. It was love at first site.

As a hard-core techie and lifelong innovator, I was embarrassed that I had put off playing with smart home devices for so long. I had been so laser focused on completing a previous book, implementing enterprise systems and developing apps that I just had not gotten around to testing smart home devices. Well, today was the day. I anxiously asked the sales associate to help me purchase a TP-Link smart light bulb and smart socket. The cost of the smart bulb was $14.97 and the socket $33.50.

By chance, my son was visiting for the weekend. He is a better technologist than I am. He had already made his apartment beyond a 'smart home' into a 'genius home', having recently installed a bunch of Nest Thermostats and most recently smart window shades! They are very energy efficient and oh so cool!

When I got home with my new smart toys we dove right into our new project.

First, we created a quick plan with some simple tasks. Later in the book I will share why you always need a plan and how to create one:
1. Install the smart bulb and turn it on and off using an app. This bulb was for our lamp behind the couch.
2. Install the smart electronic socket so we could turn a 'dumb' device, which had regular light bulbs (no smart bulbs) on and off from an app. This lamp was behind a chair in the middle of the living room.
3. Turn the smart bulb and dumb lamp on and off using voice commands, as an option to using the app
4. Create more complex scenarios where we could control both lamps at the same time using one voice command
5. Have the smart devices send us a customized email when the lights were turned on and off. This also proved useful for a business application I had in mind.

We challenged ourselves to do all the above in under an hour. Cool, right?

There was a lot on the line, such as our 'tech reps', comfort of my home, energy savings, being green and potentially some

very big business benefits, which I will get to later in the story. Think of this as a valuable fun experiment and proof of concept with a pot-of-gold at the end.

Using our well thought out project plan that took 5 minutes to create, we were ready to execute the steps. Project plans, even on the professional level, should be comprehensive but not overly complicated.

First, we unscrewed the regular light bulb from the lamp behind the couch and replaced it with a TP-Link smart bulb.

The bulb we used contained some miniature components making it a self-contained IOT device. Not all smart bulbs are fully contained; some need complementary components to work. The bulb communicated with my wireless network and made itself a known object. That was easy.

We then installed the smart socket. It has a compact design so it only uses one of the two outlets on the wall socket. The other socket can still be used for other purposes. The wireless network also saw this device and made it available to us. Very cool.

We then installed the TP-Link Kasa app and synched it to the smart bulb and the smart socket. We could see the bulb and the socket in the app. The Kasa app is very user friendly and only took a couple of minutes to install.

The app has many cool and simple to use controls. We tested turning devices on and off and it worked like a charm. Later we played with scheduling on and off for a future time. We marveled at the metrics the app provided. Up until now I envied Jeff Emmet, the former CEO for GE who was a big proponent of going digital. The millions of data points his GE Jet engines delivered in real-time allowing them to convey maintenance and performance opportunities was valuable in terms of cost savings and customer experience by reducing down-time. Well, David and I had the smart little light bulb, 'that could'.

To summarize, we installed the smart devices, installed the app and tested our ability to control the devices from anywhere in the world. An immediate benefit for me is that I no longer had to leave my home lights on all day for my dog. I could simply turn them on from the app when it got dark. In fact, I did just that from Cheesecake Factory that evening. How is that for rapid return on investment (ROI)!

Everything we accomplished to this point took us only 25 minutes. Onward we go, there's more fun and payoff ahead...

Our next milestone (a fancy project management term for reaching a major project event) was to command the smart devices from my digital assistant Google Home. We fired up the Google Home app and connected with TP-Link by indicating a few simple settings. It took approximately 3 minutes to set it up.

So, now we could say a verbal command to Google Home, such as 'Hey Google, turn on the couch light' and it would turn on-or-off the proper device or multiple devices.

Next, we wanted to get a little fancy. We wanted to the system to email us when it was instructed to turn on or off the lights. David told me about IFTTT (If This Then That). This platform is a game changer. I classify it as a new breed of simple to use data-bridge software that opens a world of possibilities.

IFTTT is a free web-based service to create chains of simple conditional statements called applets. It is the glue between systems and devices. What it is doing is greatly simplifying using powerful Application Programming Interfaces (API's) so anyone can benefit from them. We discussed value of API's in the previous chapter, 'Rocking the World with Digital Transformation and Convergence' and will again later in the book in more detail. They as one of the most valuable building blocks on the road to digital disruption.

An IFTTT applet is triggered by changes that occur within other web services such as Gmail, Facebook, LinkedIn, Instagram, Pinterest or in smart physical devices. If something happens in one of these services (If This) a trigger action can occur in IFTTT and that can extend to triggering additional actions (Then That). For our purposes, we communicated with TP-Link, which then communicated with the smart physical devices. By 'programming' simple intuitive instructions in IFTTT magic happens.

For example, using a few simple statements in IFTTT empowered us to say to Google Home: 'Hey Google, I am home.' Google Home would say 'Welcome home, I hope you had a nice day.' It would then turn on the smart bulb and the dumb lamp using the smart socket. IFTTT would send us an email about the event.

I also programmed IFTTT so my wife can close all the lights by saying one command, 'Hey Google, good night'. My wife is a smart home proponent now. She wants smart thermostats, smoke alarms and maybe even those cool smart window shades.

I decided to test another IFTTT trigger without using Google Home as part of the process. I simply copied the applet and changed the initial trigger to an email received with a keyword in the subject line rather than a using a voice command. It was easy to make the change and it worked!

We had tons of fun and completed the entire implementation in less than an hour, so our 'tech reps' are secure.

But wait! There is bonus IoT business value ahead!

As a digital technologist and business continuity director for a large corporation, one of my specialties is implementing cutting edge digital situational awareness tools and intelligent mass notification systems. These tools can save lives and businesses. I have even used them to increase revenue. They provide benefits not possible with manual tools.

Mass notification tools such as Everbridge (everbridge.com) have various ways to take input such as through an API and an easier feature called email ingestion. For example, hooking into a physical alarm system can be lifesaving in an active shooter situation. Sending real-time alerts can save product stored in warehouse by monitoring humidity and sending an alert to a group when a threshold is exceeded. Alerts in modern mass notification systems can be sent by text, email, voice, push to an app and even activate a desktop pop-up, when every second counts. Thousands of consistent messages can be delivered quickly with the push of a button or as you will see below, without even pushing a button. I have found from experience, that when such a tool is brought into an organization there is pent up demand from many people for interesting new uses.

After implementing my smart home and having the opportunity to play with some off-the-shelf tools and free platforms such as IFTTT the proverbial light bulb went off in my head, as it often does. Smart light bulbs, thermostats, weather stations and many other IOT devices can be very valuable to any type of organization. Smart devices allow us to gather important data which was never possible before.

I decided to do a demo for my management demonstrating how my little smart light bulb project could be extended to commercial applications. The demo was simple and consisted of turning on a light and passing variables such as time, date and level of brightness to IFTTT.

IFTTT would then pass the information to Everbridge. Everbridge would then automatically analyze the variables. Depending on the data and thresholds it would initiate a communication to the proper group of people that would benefit from the information.

This entire event occurred from start to finish with zero human involvement, although I could have added it in case we wanted someone to check it before sending the Everbridge generated communications. Imagine how valuable this automated process could be in a fast moving hazardous man-made or weather event where seconds count and you do not want people encountering any danger. You could even push silent alarms out in the case of an active shooter scenario.

Management loved the demo and quickly realized benefits using many different types of IOT devices. We are now working on implementing our ideas and in some cases tying them to our mass notification system and other systems. I feel like a kid in a candy store. Sometimes I must pinch myself that I get to play with this futuristic digital technology and get paid for it.

Let your imagination go. New IoT devices are coming to market at a rapid rate. If you are creative you will find the right combination of IOT devices and software to build amazing solutions internally or for the delight of your customers.

In summary, the best way to get comfortable with 'what is possible' is to play with a variety of technologies and build something small. You can then turn that into massive success.
Have Fun!

Step 2

Understanding the New Technologies

In this section I want to get your idea juices flowing! The goal is for you to be able to understand the value of each technology and start formulating ideas how you can use it in your business. You will become comfortable and knowledgeable with each emerging technology in simple and straightforward terms. This chapter is not intended to be an exhausting deep-dive highly technical manifesto.

I will describe the current and future value of each new technology. I will also share information on companies that are leading the way to making these technologies realities.

The emerging technologies described in the upcoming chapters allow you to build products and deliver services that are better AND cheaper AND faster! Never before has that equation been possible. They make the impossible possible. They enable you to transform your business and career by offering new products or enhance existing ones.

As you read about each of the emerging technologies think about:
- How you can use any or all of these technologies to enhance the value of your product or service
- How you can use any or all of these technologies to create new products or services that create value far beyond what the competition offers
- How you can combine technologies using Application Programming interfaces (API's) to create unique new customer delighting products and services

For your convenience, up-to-date links to all companies and products discussed in this section plus new golden digital nuggets are listed on DigitalSuccessBook.com/links.

The Internet of Everything

The Internet of Things (IoT) will soon be the Internet of *Every*thing. It will create immense opportunities across every facet of society.

Interestingly, The Internet of Things is no one 'Thing'. It is a fantastic concept. IoT enables us to connect digital systems with billions of physical devices. All the emerging exponential technologies we will discuss in the following chapters of this section of the book can be part of an IoT solution.

IoT consists of physical objects, sensors, networks, systems, application programming interfaces (API's) and simple data-connectors. IoT devices can monitor every type of event, in real time. Many solutions use off-the-shelf components combined in creative ways that create customer delight.

IoT leverages exponential microprocessor improvements (Moore's Law), increased bandwidth and the rapid growth of networks. These factors in combination empower you to create solutions that were impossible only a few years ago.

Connecting IoT devices, systems and networks does not take complex programming on your part. Often, there is no programming required. It is easy, fun and will get easier as we move into the future.

IoT connectivity is being integrated into industrial processes, transportation routes, workforce practices, buildings and other operational systems. It is improving and revolutionizing the efficiency, productivity, and effectiveness of the way people, businesses, schools and governments are conducting a growing variety of tasks and responsibilities. In short, it is and will be used everywhere.

IoT devices can be connected to produce torrents of real-time big data and metrics. This information can be analyzed and turned into insights. Manual labor can be taken out of the equation for many products, services and processes. Data is the food for artificial intelligence systems, quantum computers, graph databases, the blockchain and many other connected systems and objects. IoT can be the cook that serves this food.

I use an IoT tracking device called SpotTrace which bounces location data off satellites every few minutes. I then access the location data with software and do anything I want with it. I can pull it into an app or use it in back-end systems as part of a larger solution, such as displaying multiple moving objects on dashboards.

Although satellite voice communication can be costly, this device is inexpensive for a full year of satellite connectivity. Imagine what you can do with one of these to track products or vehicles. It is the size of a deck of playing cards - 2"x3".

I use IoT solutions that report temperature, motion and geo-positioning of mobile assets. Many incorporate low cost, off-the-shelf, sensors that often cost less than $10. These IoT devices can be placed in desolate and dangerous areas. They can be coupled with a Raspberry Pi / Arduino low cost microprocessor ($5 - $40) and a tiny solar panel so there would be no need for an electrical outlet or batteries. This type of self-powering solution can provide you with critical information from anywhere in the world. Use your imagination where you can put these devices. Incorporating these types of sensors in your products can bring them to life.

Placed in a food or beverage warehouse, inexpensive sensors can provide great value. They can integrate with notification tools to monitor when humidity or temperature is above a user defined threshold level. This enables the right people to know in real-time that there is an issue that needs to be fixed quickly.

Another valuable industrial use of IoT is monitoring conveyor lines for threats, such as anthrax or costly mechanical breakdowns. Certain sensors can identify packages on a conveyor that emit a dangerous odor or display a visual anomaly.

Companies that have hundreds or thousands of locations can save money and reduce compliance issues using IoT. Safety devices such as fire extinguishers, must be checked on a regular basis for pressure. The industrial age way to do this is to have a human or many humans manually check each of the devices to insure each device is in good working condition. Manually this can be very costly, error prone and slow.

All of that can lead the device to not being available when it is needed the most - when lives are on the line. Imagine if the device did not work during a fire. There can also be significant OSHA compliance violations which can be very costly, as one violation builds on the fines of the previous violation(s). Each time you are non-compliant the fine increases. If the violations are serious and happen often, OSHA can close your business until it becomes compliant. You do not want that to occur.

Fitness trackers are another popular use of IoT. They can measure the number of steps and pace you run or walk. Some can also measure heart rate, pulse rate and other important vital signs. They use inexpensive off-the-shelf components such as accelerometers, GPS and blue-tooth. The information they produce can be lifesaving or in my case play a big part of winning a national racewalking championship. My family and friends participate in a friendly online group that tracks our steps. I love my fitness tracker and keep it busy with 23 million + steps as of this writing.

Did you know that IoT devices can still provide great value without accessing the Internet? Even devices that must ultimately communicate data to a central source can continue to gather data without connectivity and then send a batch of data when the Internet is available. For example, my Fitbit tracker uses Bluetooth to communicate with my smartphone and does not need the Internet to monitor steps.

Most people are not aware that smart home devices such as window shades and track lighting can use short digital signals to communicate between devices using inexpensive Zigbee type devices. Zigbees natively have a range of 10-100 meters line of sight due to the low power consumption of the devices but by using mesh networks they can communicate over long distances. We will discuss mesh networks, which I think will grow exponentially, in the upcoming chapter, 'Smart Digital Communications - Tools & Technologies'. The Zigbee Alliance (Zigbee.org) is a good source for more information and includes a listing of lots of cool products.

Beacons are another technology that enables local communications that I predict will grow exponentially in the near future. Beacons are already being used by retailers and the supply chain to provide location based services and contextually relevant messages. Kontact.io is a good information source for information and practical uses of beacons and Bluetooth.

Sensors the Bridge Between Physical & Digital Worlds

Sensors are a key component of IoT devices. They are the bridge between the physical and digital world. Sensors respond to events in the physical environment and convert the analog information that we and others see, hear and feel into digital information consisting of bits and bytes. Once information is in digital format it can be understood and enhanced by computers. The value of the raw data can increase significantly. It can be applied to saving lives, increasing revenue and reducing expenses. It can allow you to identify opportunities and threats before the competition.

Sensors measure or receive stimulus in the form of light, temperature, pressure, sound, radiation level and then convert that stimulus into an electronic signal and transmit the signal to a measuring or control instrument. Remote sensors allow us to study an object without coming into direct contact with it. For some use cases that is a game changer. Imagine sensors that measure the intensity of storms or are encapsulated in small inexpensive CubeSat satellites or travel to the bottom of the ocean.

When your product can understand the environment and collect data, it opens up a whole new world of customer value and analytics for your company. You may even find value attaching sensors to your current products to make them smart. Smart companies are adding sensors to baseball bats, tennis rackets, smart pills, security robots. Billions of devices are now connected to the Internet and by 2025 there will be over a trillion connected devices.

Data collected by IoT devices can be analyzed locally within the device if it has a microprocessor and software or is sent to and analyzed in the cloud. Analysis can mean anything from using simple algorithms or applying artificial intelligence to provide more advanced insight.

Imagine how valuable an IoT device coupled with machine vision can be to identify a gun that enters a school perimeter. An alert can be sent to school authorities BEFORE the shooter is in the school. Once the shooter in the school there will be casualties. Smart municipalities are already applying digital technology to this type of problem and it will save lives.

The technology of modern remote sensing began with the invention of the camera more than 150 years ago. Although the first, rather primitive photographs were taken as "stills" on the ground, the idea and practice of looking down at the Earth's surface emerged in the 1840s when pictures were taken from cameras secured to tethered balloons for purposes of topographic mapping. Today satellite images of parking lots during holiday time can tip off investors of how stores are performing - and yes, some innovative investors are already doing this.

Perhaps the most novel application of sensors at the end of the last century is the famed pigeon fleet that operated as a novelty in Europe. Fast-forward to World War I and cameras were mounted on airplanes that provided aerial views of large surface areas. Although these could not be downloaded in real-time they proved invaluable in military reconnaissance. Aerial photography remained the single standard tool for depicting the surface from a vertical or oblique perspective until the 1960's.

Modern uses for sensors can be counted in the millions including monitoring the performance of jet engines in flight, truck telematics safety and proactive maintenance metrics, levels of gas in fuel tanks, smoke and carbon monoxide, noise, traffic, air pollution, earthquakes, radiation, wind and inventory product levels. The variety of inputs that can be measured is huge!

Remote sensing instruments are of two primary types—active and passive. Active sensors provide their own source of energy to illuminate the objects they observe. Passive sensors, on the other hand, detect natural energy (radiation) that is emitted or reflected by the object or scene being observed. Reflected sunlight is the most common source of radiation measured by passive sensors.

RFID can be a good solution for tracking products through the supply chain or as inventory in a store. The tags are tiny, light and inexpensive. They can be applied to most any product. I worked on one project that used active sensors to monitor the location of jewelry.

Many types of sensors cost a fraction of what they did only a few years ago. Some, that formerly cost $25-$100, now cost only pennies. This opens many new use cases. Eventually every product will use a sensor for added value.

I envision sensors being integrated into throwaway umbrellas that will send rain data to servers in the cloud. That will enable predicting weather to very specific points more accurately than current weather stations. Accurate weather predictions can have significant commercial value. The umbrellas could even be given away for free, as the collected data will provide much greater value than the cost of an umbrella.

I also envision sensors being embedded in our skin. Oh wait, I am too late on that one, as it is already being done. There is at least one business that provides employees the option to embed sensors in their arm. It may surprise you that most people, given the option, wanted the implant. It provides value in eliminating daily manual processes.

If you take my earlier suggestion about playing with digital technology for success, knowledge and fun, there are many online stores that sell sensors. Sensors can inexpensively be coupled with tiny microcomputers, and small radio transmitters to create useful and unique new products. I have purchased cool inexpensive components from Sparkfun, Adafruit and Electronics Hub. The all have a wide variety of sensors and many other Maker products.

To get your ideas flowing here is a small sampling of the types of sensors available:
- Motion
- Temperature
- Humidity
- Wind
- Vibrations
- Passive Infra-Red (PIR)
- Ultrasonic

- Pressure
- Light
- Location
- Height
- Fuel
- Fluid Velocity
- Touch
- Color
- Accelerometer
- Smoke
- Tilt
- Gas
- Alcohol
- Piezoelectric (measures changes in pressure, acceleration, temperature, strain, or force by converting them to an electrical charge).

I cherry-picked two ideas from NASA to give you an idea of what is possible. NASA has over 100 additional sensor applications that can be licensed their web site. If you want to review them, please visit: https://technology.nasa.gov/sensors/1

Wireless Sensor for Pharmaceutical Packaging and Monitoring Applications

NASA's Langley Research Center researchers have developed a wireless, open-circuit SansEC [Sans Electrical Connections] sensor that can be used for pharmaceutical applications without the need for physical contact. Many attributes of a container can be monitored, such as liquid or powder levels, temperature of contents, and changes caused by spoilage. Tampering can also be detected.

Benefits

- The sensor can detect damage to the package or container
- One sensor can be used for multiple measurements, including biological decay and temperature
- The sensor receives power wirelessly, eliminating the need for a sensor power source connection

- The sensor sends signals wirelessly to the data acquisition device, eliminating signal wiring
- The sensor is a single electrical component. There are no wires and no electrical connections
- The sensor can be mass produced and well suited for manufacture to a specific size
- The sensor can operate external to a package or container, protecting it from damage by environmental elements

Applications
- Package tampering detection
- Medicine dosage monitoring
- Food/ pharmaceutical spoilage detection
- Pharmaceutical/ food container level monitoring

Impact and Trajectory Detection System
  The sensor pinpoints location of impacts and trajectory of the projectile using piezoelectric polymer film at the NASA Johnson Space Center (JSC) and has developed an Impact and Trajectory Detection System that can determine the time and location of the projectile impact as well as the trajectory of the projectile.

Benefits
- Accurate: Features reliable impact and trajectory detection on a wide variety of surfaces
- Customizable: Offers protection to small or large areas
- Low power requirements: Requires minimal power for the integrated panel circuits, the external digital processor and display unit
- Adaptable: Functions over wide environmental extremes of temperature and pressure, including in a vacuum
- Adjustable: Allows for remote monitoring

Applications
- Satellites
- Spacecrafts
- Tanks and Military Vehicles

- Cargo Containers
- Storage Tanks and Containers
- Weather Stations
- Building Construction
- Aircraft
- Security Systems
- Protection Systems

API's and Simple Connectors
'The Sum is Greater than the Parts'
IoT devices coupled with application programming interfaces (API's) are the great enablers. I call them 'the secret sauce'.
API's are the software connectors that allow us to mesh physical devices, systems and big data into a 'sum is greater than the part building blocks' opportunity to identify opportunities to create extraordinary customer value. The only limits are our imaginations. ProgrammableWeb.com has a listing of 19,949 APIs that enable you to mix and match data sets. Data.gov has thousands of API's to access interesting government data sets.
Google maps has one of the most popular API's. Developers can use it to map longitude and latitude for any asset. For example, using it you can visually display all your buildings or people on a map. You can even add additional visual layers on top of a map. This can alert you to possible weather or man-made disruptive events. I use this type of visual benefit every day.
When you speak with physical IoT device or software solution providers, always ask about their roadmap to the future. Ask how they currently integrate with other systems and hardware components. Request use-cases of how their current clients are creatively using their IoT products. Ask if they offer an API or a connector to other applications.
API's come in a variety of flavors. The most popular is the simple, lightweight and fast RESTful API. If you can write excel macros or JavaScript you can handle API's, if you have the time and inclination. They are moderately easy to use, especially if you use friendly tools such as Postman or Swagger.
RESTful API's resources separate data from the presentation layer. That enables the data to be accessed and passed between destinations using many formats

including HTML, XML or JSON (JavaScript object notation). I favor JSON for the most flexibility and control.

Data Connectors

Data connectors are a recent class of software enablers that provide great value and will grow quickly as the general population understands their value. They make using APIs as easy as filling out a few pieces of information on a form and snapping systems and physical devices together. It is magic and no technical skills are required to do it. How cool is that!

Two of the major data connector platforms are IFTTT (ifttt.com), which I discussed using in the chapter, 'How to Digitally Play, Prototype and Profit', and Zapier (zapier.com). I use them both and they have great value.

Data connectors are far easier to implement than API's, as the software already embeds the API commands in pre-build connectors. You simply pick and choose what actions you want to occur and when.

There may be some cases where the data connector tool does not offer all the options you need or there may not be a pre-built connector between the systems you are using. In those cases, you can use the native API or build a connector for IFTTT or Zapier and share it with other users. The IFTTT platform also lists hundreds of public Applets other people built that you can use and modify for your own scenarios.

IoT can turn an organization from mediocre into a world-changing disruptor. IoT will be key to reinventing our world.

Next, we will explore amazing new technologies that you can integrate with IoT or use standalone to create incredible customer value externally or internally.

Artificial Intelligence(AI)
The Future is Here and It is Bright!

"There is no reason and no way that a human mind can keep up with an artificially intelligent machine by 2035."
Gray Scott, Futurist, Techno-Philosopher

"Some people call this artificial intelligence, but the reality is this technology will enhance us. So instead of artificial intelligence, I think we'll augment our intelligence."
Ginni Rometty, President and CEO IBM

"Artificial intelligence would be the ultimate version of Google. The ultimate search engine that would understand everything on the web. It would understand exactly what you wanted, and it would give you the right thing. We're nowhere near doing that now. However, we can get incrementally closer to that, and that is basically what we work on."
Larry Page, CEO Alphabet Inc.

I love all our emerging exponential technologies but If I were to pick one with the most potential to change every aspect of humanity in the shortest period of time it would be AI. It is here, it is now and it is part of your life, whether you know it or not. Some of the things I share in this chapter will light you up to the possibilities and perhaps change the course of your business and personal future.

For the reasons above this is the longest chapter in the book. In fact, I could have easily made this chapter into a book. With your encouragement, perhaps I will write that book.

In this chapter, we will discuss:
- What AI is and its value to you
- A brief history of AI
- Amazing AI advancements and successes
- Tools that can help non-data scientists play with AI and even develop applications

As always, think about how you can leverage this amazing emerging technology for your current or future business and career in new innovative ways. I will provide ideas as well.

AI Overview

AI is the ability of machines to think, learn and be creative. Yes creative! I describe a story in this chapter where experts feel an advanced deep learning machine has already acted creatively.

Most likely AI will change your industry, profession and personal life. Now is the perfect time to begin learning how you can use it to your advantage.

AI is being used in autonomous vehicles and drones to make critical real-time decisions. Retailers will soon be loading products on trucks before you even order them. Yes, they will know what you need and when you need it before you do. Financial Institutions can detect fraud using powerful algorithms. Professions in fields diverse as medical, legal and customer service are becoming better and faster using AI. Later in the chapter, 'How to Leverage Content, the Currency of the Digital World', you will learn about an innovative company that produces automated news stories and baseball box scores that are indistinguishable from human reporters using AI and big data.

Combining artificial intelligence with big data and smart algorithms is the secret sauce. It can deliver predictive analytics that help you solve complex problems and even help to predict the future. AI can comb through terabytes of data in seconds and identify hidden patterns. AI can accurately predict what will happen next, whether it is a threat such as a tornado, an active shooter planning an attack or opportunities such as how to medically treat a patient or the likelihood of an increase in a stock price.

AI and big data also raise the bar on preventive maintenance of industrial machines. GE pumps terabytes of data from a single jet engine on a single flight in real time through AI systems to understand where and when maintenance should occur. Compare this to waiting for a scheduled maintenance or worse, until an engine fails. Every airline should be doing this. Casualties are no longer acceptable because of parts breaking from fatigue and possible catastrophic casualties.

Telematics on trucks can monitor, in real time, critical engine and tire health and driving patterns by the operator. Telematics improves safety and reduce expenses. Tires can communicate that they need replacing, engines can send alerts to fill fluids or impending engine trouble. The alternative is to wait until a tire blows or an engine fails.

Telematics can also suggest when to reduce or increase speed. It can pinpoint if a driver is hitting the breaks to often or too hard. Maybe a larger distance from the vehicle in front of them can help. The number of scenarios AI and big data can help operators is wide and valuable.

Smart home assistants including Amazon Alexa and Google Home use artificial intelligence. If you are not experimenting and enjoying one of these devices yet, I am confident one will be in your near future. My wife is not a 'technical' person but she gets so much enjoyment from Google Home. In addition to making our lives easier our Google Home device has allowed me to experiment by integrating voice commands with smart light bulbs and electrical sockets and run them through a data bridge, IFTTT, to my organizations mass notification system.

AI infused chatbot's are becoming extremely popular. They benefit customers 24x7 and reduce expenses for the organizations employing them. Now is a great time for companies to begin understanding the business value in chatbots. Facebook messenger is a platform that has launched over 100,000 different chatbots.

Short for "chat robot," a chatbot is a computer program that simulates human conversation, or chat, through artificial intelligence. ALICE (Artificial Linguistic Intelligent Computer Entity) is an open source, natural language chatbot. It uses artificial intelligence for human interaction.

Chatbots are beginning to get personalities and the ability to conduct increasingly sophisticated conversations. Retail companies use them to intelligently answer questions from online shoppers. Most any company can benefit from a smart chatbot.

The National Cancer Institute (NCI) is exploring how chatbots can be used in improving health.

AI encompasses Machine Learning and Neural Networks which take traditional rule based AI to a breakthrough level. Combining AI with other emerging technologies, such as robotics, empowers us to create machines that think.

AI can provide answers, improve predictions and solve problems. It does certain tasks thousands and millions of times faster than a human can do them. It can power robots to do dangerous and tedious tasks with close to zero errors.

Three types of AI learning can occur:
1. Supervised learning in which the machine analyzes high quality historical data and makes decisions about future data
2. Unsupervised learning in which the machine makes inferences about future data based on patterns it finds within past data
3. A combination of supervised and unsupervised learning

The U.S. Government is as excited about AI as I am. In a 2018 strategic plan, the Government Accountability Office (GAO) identified AI as one of the "five emerging technologies that will potentially transform society." What we have seen from our unique government-wide perch at GSA is that agencies are tremendously interested in AI and other emerging technologies - not because they are the latest fad, but because people recognize the potential to transform and simplify the way Americans interact with their government."

A March 2018 hearing, 'Game Changers: Artificial Intelligence' detailed four ways the U.S. Government is supporting AI evaluation and adoption:
1. The Federal Acquisition Service (FAS) provides contracting vehicles and mechanisms including Schedule 70 and its associated programs, Startup Springboard and FAStlane, as well as several other government-wide acquisition contracts, which encourage competition and help connect agencies and businesses to allow government to efficiently procure the most effective new AI services.
2. We are piloting within our agency Robotic Process Automation (RPA) and other related technologies that are designed to augment our workforce to achieve more with less and establish a foundation for greater data-driven decision-making through AI.

3. The interagency Emerging Citizen Technology Office (ECTO), we are helping support and coordinate government-wide development of citizen-facing AI programs, both public-facing as well as for internal agency use, with active participation from both the public and private sectors.
4. Along with our private sector and federal agency partners, we are pursuing a greater understanding and alignment of IT modernization through cloud adoption, data services, and emerging technologies, including AI, that deliver the greatest benefit to the American people.'

The Department of Homeland Security has stated they believe that the future AI trajectory will proceed in the following three ways:

1. First, AI technology is increasingly providing us with new knowledge and influencing action. Fueled by sensors, data digitization, and ever-increasing connectedness, AI filters, associates, prioritizes, classifies, measures, and predicts outcomes, allowing the Federal government to make more informed, data-driven decisions.
2. Second, algorithms are ingesting and processing ever higher volumes of data. Their complexity, especially in the case of deep learning algorithms, will continue to increase, and we need to better understand how outputs are produced from the set of inputs, which may not be able to be understood or analyzed in isolation.
3. Finally, private industry is leading the way in AI development, as many see the implementation of AI as a key competitive advantage. The private sector's significant investments and the ability to adopt new AI models and processes faster than the public sector present the government with a key decision point on how to best participate in this growing, but still nascent field.

Government should move forward with adoption of emerging technologies such as AI to improve citizen services. Government plays an important role in promoting research and development. Government should ensure it is informed of developments in the private sector, while continuing to support AI research and development, and promote the use of AI technology to create government efficiencies and enhance the public good.

AI - A Brief History

To give you the full story of how we got to the present and how fast improvements in AI are occurring let us take a whirlwind AI history tour.

Artificial Intelligence started out with a bang in the 1940's. The earliest modern successes in AI used if-then branching called search trees or reasoning as search. Programmers had to provide, at a minimum, every question and answer. This proved limiting, tedious and often impossible, as many subjects are so broad and have so many data points that it would be impossible to input all the information and keep it up to date, although there were some valiant attempts to create all-encompassing databases.

In a way, the early quick 'successes' of search tree logic were unfortunate as they proved to be fool's gold. Simple programs could be created using 100% branching rules to figure out the correct answers but attempts at creating more meaningful solutions proved futile. They did not have enough flexibility and could not scale. Even if search trees had achieved more success they would have eventually hit a brick wall.

Unfortunately, based on the early successes of the search tree method, overly optimistic predictions were prematurely made that AI would rival our brains in only a few short years. There were some valiant attempts at creating general artificial intelligence machines through the early years, unfortunately, due to the technology available at the time and the faulty approach, many did not work.

In the early 1970's progress slowed and the lofty predictions for AI that did not materialize as quickly as anticipated caused AI development to grind to almost a halt.

In 1983, or thereabouts, there was a computer / AI hands on demonstration at the Philadelphia Science Museum. I had read that attendees would have an opportunity to try Eliza, which was an early AI program.

For a budding technologist like myself, this was too exciting to miss. I had to be there. I piled my wife and in-laws into my car and made the drive from New York to Philadelphia. They were good sports about it and enjoyed the show or so they told me.

I typed in questions and Eliza would provide answers. I enjoyed the interaction but some of her answers to simple questions were often very wrong and sometimes hilarious. I can't say Eliza was disappointing, as she was the best at the time. Even IBM's Watson many years later had some funny responses when playing in the famous Jeopardy matches against Ken Jennings and Brad Rutter. Today Siri, Google Home, Alexa, etc. can provide some comical answers to certain questions, but they also provide great value which gets better every day as they learn.

The 1980's ushered in renewed enthusiasm for AI due to some successes from the machine learning developers. Machine learning took a very different statistical approach at achieving true artificial intelligence as opposed to search trees. Machine learning uses data to understand patterns and make predictions.

Machine learning and neural networks continue to be the dominant focus by AI developers today. Recent achievements have been astounding. Companies such as DeepMind, which is now part of Google and IBM's Watson (both of which you will read about below) are leaders in making the exciting future of AI a reality. Many other companies are joining the AI party. AI is such a broad area with unlimited opportunity. There is a huge need for talented entrepreneurs and AI software developers and engineers.

AI is beginning to pay off in big ways. Deep learning machines are now learning on their own with a minimum of programming. This has allowed them to do things that were never thought possible, such as 'read' millions of medical journals in minutes and make suggestions to doctors that can lead to more accurate diagnosis and treatment. There are many other use cases.

AI has come a long way in a relatively short period since my days with Eliza. I attribute this growth to Moore's law, innovators and brilliant AI leaders and developers.

Happily, technology will continue to advance exponentially which will make our current astonishing AI achievements seem basic. When quantum computing kicks-in there will be a 'quantum-leap' in AI capabilities to simulate the brain. I predict computers will achieve parity with the human brain before 2030. In 2018 we have already created AI machines on par with a mouse, which is quite impressive.

Now, let us make a quantum-jump from Eliza to DeepMind's AlphaGo Zero. You will see why I am confident the best is yet to come and it will drastically change our world.

The Amazing Story of AlphaGo Zero

The story that follows may seem like science fiction but I assure you it is not. It is a perfect example of the power of artificial intelligence and how exponentially fast it is progressing. It is a story that will help you see what is possible using AI - which is anything you want it to be.

I attribute much of the story of AlphaGo and AlphaGo Zero to my friends at Google DeepMind who permitted me to reproduce content from their website along with my insights and experience in this chapter.

What is Go?

The game of Go was invented in China over 3,000 years ago. The rules are simple, players take turns to place black or white stones on a board, trying to capture the opponent's stones or surround empty space to make points of territory. Even though the rules are simple, Go is very complex. There are an astonishing 10 to the power of 170 possible board configurations. That is more than the number of atoms in the universe. It is far more complex than Chess.

Mastering the game of Go

Go has long been viewed as the most challenging of classical games for artificial intelligence. Despite decades of work, the most sophisticated computer Go programs were only able to play at the level of human amateurs. In fact, although I anticipated AI computers beating chess grandmasters and I wrote simple 'AI' rules programs to beat tic-tac-toe (don't laugh, it was a great way to learn programming) and checkers, I did not think I would see a computer proficient at Go in my lifetime. Never underestimate innovative people and exponential technology.

Traditional AI methods, such as my little tic-tac-toe program, which construct a search tree over all possible positions, don't have a chance in Go. The incredible number of possible moves and the difficulty of evaluating the strength of each possible board position make the rules approach impossible.

To capture the intuitive aspect of the game, DeepMind knew that they would need to take an innovative approach. AlphaGo combines an advanced tree search with deep neural networks. These neural networks are key, as they take a description of the Go board as an input and process it through a number of different network layers containing millions of neuron-like connections. One neural network, the "policy network", selects the next move to play. The other neural network, the "value network", predicts the winner of the game.

DeepMind showed AlphaGo many strong amateur games to help it develop its own understanding of what reasonable human play looks like. Then they had it play against different versions of itself thousands of times, each time learning from its mistakes and incrementally improving until it became immensely strong, through a process known as reinforcement learning.

AlphaGo became the first computer program to defeat a professional human Go player, the first program to defeat a Go world champion, and for a time it was arguably the strongest Go player in history (more on how that changed below).

AlphaGo's first formal match was against the reigning 3-times European Champion, Mr. Fan Hui, in October 2015. Its 5-0 win was the first ever against a Go professional, and the results were published in full technical detail in the international journal, Nature. This was a great accomplishment considering that Go is not as popular in Europe as it is in Asia.

AlphaGo then upped the ante and went on to compete against legendary player Mr. Lee Sedol, winner of 18 world titles and widely considered to be the greatest player of the past decade.

AlphaGo's 4-1 victory in Seoul, South Korea, in March 2016 was watched by over 200 million people worldwide. It was a landmark achievement that experts agreed was a decade ahead of its time, and earned AlphaGo a 9 dan professional ranking (the highest certification) - the first time a computer Go player had ever received the accolade.

Perhaps, a machine can be quite creative. During the games, AlphaGo played a handful of highly inventive winning moves. Several, such as move 37 in game two, were so surprising they overturned hundreds of years of perceived wisdom, and have since been examined extensively by players of all levels. In the process of winning, AlphaGo taught the world completely new knowledge about what appears to be the most contemplated game in history.

Since then, AlphaGo has continued to surprise and amaze. In January 2017, an improved AlphaGo version was revealed as the online player "Master" which achieved60 straight wins in online fast time-control games against top international Go players.

In May 2017, Alpha Go took part in The Future of Go Summit in the birthplace of Go, China, to delve deeper into the mysteries of Go in a spirit of mutual collaboration with the country's top players.

If you think the story is fantastic so far, read on...it gets even better!

AlphaGo Zero: starting from scratch

In October 2017, DeepMinds' AlphaGo Zero paper was published in the journal Nature. Unlike the earlier versions of AlphaGo which trained on thousands of human amateur and professional games to learn how to play the game, AlphaGo Zero bypasses this process and learns to play the game of Go without human data, simply by playing games against itself, starting from completely random play.

Over the course of millions of automated games played in lightning speed against itself with no human intervention, the system progressively learned the game of Go from scratch, accumulating thousands of years of human knowledge during a period of just a few days! AlphaGo Zero also discovered new knowledge, developing unconventional strategies and creative new moves that echoed and surpassed the novel techniques it played in the games against Lee Sedol and Ke Jie.

These moments of creativity suggest that AI will be a multiplier for human ingenuity, helping DeepMind solve some of the most important challenges humanity is facing.

After just three days of self-play training, AlphaGo Zero emphatically defeated the previously published version of AlphaGo- which had it self-defeated 18-time world champion Lee Sedol- by 100 games to 0. After 40 days of self-training, AlphaGo Zero became even stronger, outperforming the version of AlphaGo known as "Master", which has defeated the world's best players and world number one player Ke Jie. This makes AlphaGo Zero possibly the best Go player ever.

Experts described the paper as "a significant step towards pure reinforcement learning in complex domains". DeepMind made this progress by streamlining the architecture behind Zero; they unite the policy and value networks into a single neural network and incorporate a simpler tree search that relies on this single neural network to evaluate positions and sample moves, without performing rollouts of the games. This can be thought of as using a single top level professional to advise the system on its next move, rather than taking a crowdsourced answer from hundreds of amateur players. The simplicity of AlphaGo Zero's architecture also dramatically speeds up the system while also lowering the amount of computer power it needs.

This technique is more powerful than previous versions of AlphaGo because it is no longer constrained by the limits of human knowledge. If similar techniques can be applied to other structured problems, such as reducing energy waste or searching for revolutionary new materials, the resulting breakthroughs have the potential to positively impact society.

IBM Watson and Chatbots

Ginni Rometty, IBM's President and CEO as I write this in August 2018, has a plan and exponential technologies are a big part of her plan. I describe a few of IBM's AI products below. IBM is also a leader in developing blockchain technology and quantum computing. Eventually, combining these exponential technologies and others will advance society to a far better and smarter place. I believe in her vision. It will happen.

IBM has been a long-time believer in the value of artificial intelligence. Perhaps you were one of millions of people who watched IBM's Watson computer beat Ken Jennings and Brad Rutter in Jeopardy in 2011. It opened a lot of eyes to what is possible, including mine.

Watson is also now a fraction of the size of its Jeopardy winning ancestor and more powerful. There is that amazing exponential technology equation again.

Watson is now being used for many commercial applications. You might think these applications are expensive but that is not the case. Many of the AI tools are well within the reach of even the smallest startups and the payback in customer delight can be impressive. For example:

Watson Assistant - I spoke of the value chatbots will provide and Watson Virtual Agent allows you to quickly and easily build a powerful chatbot for your organization. No complex machine learning experience is required. The chatbot offers a cognitive, conversational self-service experience that can provide answers. It is highly customizable to fit your business. It also provides deep analytic insights on your customer's engagement with the chatbot to help you understand dynamically your customer's needs - https://www.ibm.com/us-en/marketplace/cognitive-customer-engagement

Watson Explorer - is a content analysis platform that lets you listen to your data for advice. You can explore and analyze structured, unstructured, internal, external and public content to uncover trends and patterns. That can help you improve decision-making, customer service and ROI. This can provide you with a holistic view of your customers to deliver what they need and will make them happy - https://www.ibm.com/products/watson-explorer

Watson Knowledge Studio - You can teach Watson to discover insights in unstructured data without writing any code - https://www.ibm.com/watson/services/knowledge-studio/

The Watson team developed a short video that demonstrates how you can build a simple chatbot in 6 minutes. I enjoyed it and as I mention throughout the book there is nothing like doing hands-on work to understand how to delight customers with technology- https://www.youtube.com/watch?v=MTCc4d-RXP0

AI Career and Development Tools
Career Alert - If I was beginning my career in technology AI is an area I would seriously consider. Developing AI is not easy, but it is becoming more accessible to people not versed in statistics and complex AI programming languages. There is an exploding need to make development in AI within reach of a wide audience of developers and our vendors are stepping up! Amazon Web Services (AWS), Microsoft Azure, IBM Cloud / Watson and Google Cloud all have powerful cool AI development tools and training.

Gluon - (https://aws.amazon.com/blogs/aws/introducing-gluon-a-new-library-for-machine-learning-from-aws-and-microsoft/) is a library for machine learning from AWS and Microsoft. It includes an open source deep learning interface. Its goal is to allow developers to easily and quickly build machine learning models, without compromising performance.
Gluon provides an easy to use API for defining machine learning models using a collection of pre-built, optimized neural network components. Developers who are new to machine learning will find this interface more familiar to traditional code. More experienced data scientists and researchers will appreciate the ability to build prototypes fast and utilize dynamic neural network graphs for new model architectures, without sacrificing training speed. Gluon is available in Apache MXNet and probably in additional frameworks as of September 2018.

TensorFlow
(https://ai.google/tools/tensorflow/#?modal_active=none) is a popular open-source software library for machine learning. It is a fast, flexible, and production-ready open source machine learning library for research and production. It supports a variety of applications, with particularly strong support for

training and inference with deep neural networks. TensorFlow operates on different devices, from smartphones to data centers, and is a machine learning platform that everyone can use.

Neo4j (neo4j.com) is a graph database that can deliver artificial intelligence capabilities and many additional benefits. One of our profiled companies in this book, devRant.com, uses neo4j as part of their software stack to deliver recommendations. A graph database can be a great way for database developers, including myself, to deliver powerful AI solutions rapidly.

Visual Studio Tools for AI (https://www.visualstudio.com/downloads/ai-tools-vs/) is an extension that supports deep learning frameworks including Microsoft Cognitive Toolkit (CNTK), Google TensorFlow, Theano, Keras, Caffe2 and more. You can use additional deep learning frameworks via the open architecture. Visual Studio Tools for AI leverages existing code support for Python, C/C++/C#, and supplies additional support for Cognitive Toolkit BrainScript. It empowers you to build, test, and deploy deep learning and AI solutions

Mahout (http://mahout.apache.org/): is a scalable machine learning and data mining library that runs on top of Apache Hadoop. It simplifies the identification of patterns in big data.

Eclipse Deeplearning4j (https://deeplearning4j.org/) is an open-source, distributed deep-learning library written for Java and Scala. It can be integrated with Hadoop, Apache Spark and DL4J to bring AI to business environments for use on distributed GPUs and CPUs.

I hope this chapter excited you to the possibilities of AI. It is improving at an astounding rate and I am sure the best is yet to come.

3D Printing
The Power to Create Anything!

When we think of 3D printers we think of the early home versions that simply printed trinkets out of plastic. Emerging technologies start that way, using simple use cases. They are often powered by the home market. Then rapid growth takes over. Finally, businesses leverage the momentum of emerging technology. It is important to be able to identify future technology trends early in the game so you are ready to pounce when opportunity appears. 3D printing is an opportunity that is here, it is real and it is presenting many new business and career opportunities!

3D printing is revolutionizing health, manufacturing, retail and design. What you are able to 3D print is only limited by your imagination. It is enabling organizations and individual's users to create custom products in new ways, reduce waste, save energy and generate revenue!

Some of the current use-cases of 3D printing boggle the mind. In fact, my editor questioned my sanity when I described 3D printing of full size cars, houses, planes, robots, custom shoes, musical instruments and more. I proved each example to her and they made it into the book!

There are many important advantages to 3D printing and some are life changing:

3D printing makes it economical to print unique custom products. Custom has a higher perceived value than mass produced identical products. In many cases I would much prefer a custom object than a generic one. Businesses can take an existing product and make a better and cheaper version using 3D printing. The result can mean new revenue streams and reduced costs.

It is easy to 3D print a complex object without first creating a mold and pouring in plastic or metal. Creating complex molds is expensive and time consuming. 3D printing can reduce those costs. Not having to build and use molds helps translate into lower energy use for 3D printing, up to 50 percent less energy for certain processes compared to conventional manufacturing processes.

Instead you can feed the 3D printer a computer design and it will print to the exact specifications. There are platforms online that have libraries of designs that other people have made. You can print parts or send the design to a co-worker half way around the world and he or she can print the identical object.

Why not brainstorm with your team about new custom products that would benefit your customers or attract new customers? Another option is creating low-cost high-value products and giving them away as incentives to purchase your existing products!
If customer demand is high, you can automate the process so customers can design custom sizes, colors or text they want on their product on your easy to use web site. They can fill out a form or build a simulation of the product by rotating it and customizing properties. They then can submit it to your database and you can pass the information to your 3D printer. Instant customer value is created with zero effort on you part!

The use of 3D printers is snowballing. As prices of printers and materials continue to plummet and the technology becomes faster and more precise, 3D printing will change the way companies and consumers think about manufacturing.
3D printing makes experimenting with new concepts and designs easy and cost effective. I speak about MVP's (Minimal Valuable Products) for success later in the book, but 3D printers extend that agile mindset of building and improving in iterative steps based on user feedback to the physical world of atoms for the first time.

3D Printers come in two flavors additive and subtractive. Each has its benefits.
1. Subtractive printing (manufacturing) starts with a block of material which is reduced according to the design specs to create an object. The result: Subtractive manufacturing can waste up to 30 pounds of material for every 1 pound of useful material in some parts, as per a finding from the Energy Department's Oak Ridge National Lab.
2. The additive method is more common. It is like a laser printer that lays down layers of ink, except 3D printers

3.   Lay down layers of raw material such as plastic and metal. Innovators are experimenting using creative materials like chocolate and other food items, wax, ceramics and bio-material like human cells.

Each new layer is attached to the previous one until the object is complete. Objects are produced from a digital 3D file, such as a computer-aided design (CAD) drawing or a Magnetic Resonance Image (MRI). Additive 3D printing is very efficient and cost effective. Waste is minimal as approximately 98 percent of the raw material is used in the finished part.

Oak Ridge National Lab is partnering to develop a new commercial additive manufacturing system that is 200 to 500 times faster and will be able to print plastic components 10 times larger than today's commercial 3D printers.

The flexibility of 3D printing allows designers to make changes easily without the need to set up additional equipment or tools. It also enables manufacturers to create devices matched to a patient's anatomy (patient-specific devices) or devices with very complex internal structures. These capabilities have sparked huge interest in 3D printing of medical devices and other products, including food, household items, and automotive parts.

If airplane and cars parts can now be 3D printed, why not specialized parts required during a disaster? For the right company or possibly a city impacted by an earthquake or tornado, that could be valuable and perhaps life-saving. If a company uses a wide variety of parts it could get very expensive to insure there are adequate supplies of all back up parts on-hand at all times. Waiting for a critical part to be shipped could be slow and expensive. It could impact revenue and lead to loss of customer confidence. I am already experimenting with 3D printed backup parts in my role as a Director of Business Continuity and Disaster Recovery for an 18.6 billion dollar company.

It would be great to have the ability to print a small, yet critical, conveyor, air conditioner or printer part, if the vendor supplied online 3D plans on their website. Even if the 3D printer costs a few thousand dollars it might be insignificant measured against the expense of downtime and not being able to create and ship products to customers.

Zortrax (zortrax.com) is a 3D printer company. They describe a 3D printed parts scenario on their web site: 'One of the Bosch factories started using Zortrax M200 printers to repair production line machines. This reduced the cost of one spare from €450 (approx. U.S. $523) to €1 (approx. U.S. $1.16) and reduced the time necessary to obtain it. This, in turn, meant shortening the process of repair. Savings through this technology have already surpassed 80 thousand euros.' I expect similar solutions to be used by more and more companies in the near future.

When Elon Musk sends a team to Mars I am confident some sort of 3D printer will be on the journey. Rather than taking every conceivable part with them the deep space pioneers will be able to pick and choose from millions of print on-demand parts. The printer might even be able to build parts for pod-type dwelling structures or the entire structure.

When NASA is researching a technology, I have found it is often a good indicator of a positive trend in the future. On December 6, 2017 NASA held 'The 3D Printing In Zero-G Technology (3D Printing in Zero-G) Experiment' demonstrating that a 3D printer works normally in space. As you know from the top of this chapter, additive 3D printers extrude streams of heated plastic, metal or other material, building layer on top of layer to create 3 dimensional objects. Testing a 3D printer using relatively low-temperature plastic feedstock on the International Space Station is the first step towards establishing an on-demand machine shop in space, a critical enabling component for deep-space crewed missions and in-space manufacturing.

They described the experiment that whether in space, manufacturing or other earth bound businesses three-dimensional printing offers a fast and inexpensive way to manufacture parts on-site and on-demand. This is a huge benefit to long-term missions with restrictions on weight and room for cargo. After testing of hardware for 3D printing on parabolic flights from Earth resulted in parts like those made on the ground, the next step was testing aboard the space station. The test included printing items designed by students and results showed that 3D printers work normally in space. This work will contribute to establishing on-demand manufacturing on long space missions and help to improve 3D printing methods on the ground.

Oak Ridge National Laboratory (ORNL) (ornl.gov/gsearch/3d%2Bprinting) is on the cutting edge of uses for 3D printing and many other technologies. I very much enjoy their site and the great work they do. This is definitely a resource you should comb through if you are exploring 3D for your business and for interesting new digital ideas.

One such ORNL use-case combines clean energy technologies into a 3D-printed building and vehicle to showcase a new approach to energy use, storage and consumption. The Additive Manufacturing Integrated Energy (AMIE) demonstration, displayed at DOE's Office of Energy Efficiency and Renewable Energy Industry Day event, is a model for energy-efficient systems that link buildings, vehicles and the grid. Notice that multiple emerging technologies are applied to solve a problem. This will be emphasized in the next section of the book as we discuss how digital products are helping create a smarter new world.

An ORNL team worked with industrial partners to manufacture and connect a natural-gas-powered hybrid electric vehicle with a solar-powered building to create an integrated energy system. Power can flow in either direction between the vehicle and building through a lab-developed wireless technology. The approach allows the car to provide supplemental power to the 210-square-foot house when the sun is not shining. You can watch an animation of the energy flow here: https://youtu.be/aflTvjudnoc. The demonstration also showcases additive manufacturing rapid prototyping potential in architecture and vehicle design; the car and house both were built using large-scale 3D printers.

The 38x12x13-foot building was designed by architecture firm Skidmore, Owings, and Merrill (SOM) through the University of Tennessee-ORNL Governor's Chair for Energy and Urbanism. It was assembled by Clayton Homes, the nation's largest builder of manufactured housing. Connecting the house to the 3D-printed vehicle demonstrates the concept of integrating two energy streams, buildings and transportation, which typically operate independently. Working together, they designed a building that innovates construction and building practices and a vehicle with a long enough range to serve as a primary power source," said ORNL's Roderick Jackson, who led the AMIE demonstration project. "Our integrated system allows you to get multiple uses out of your vehicle."

I would never forgive myself if I did not 'steer' you to this amazing video of how ORNL printed a full-size car. This is the one I showed to my doubting editor. It shows the process in a time lapse and the 'way cool car' is one I could only wish I owned.

Apis Cor (apis-cor.com) is a San Francisco startup that prints buildings. Their special mission is to help people around the world improve their living conditions. Too many people world-wide cannot afford to buy or build a house. Building the industrial way is expensive and time consuming.

Apis Cor has radically changed the home building equation using innovation and powerful technology. They developed 3D printing technology, new building materials and a mobile 3D printer to build affordable, eco-friendly houses within a single day, capable of lasting up to 175 years. They have ambitious plans to print houses on every continent - even Antarctica if needed. They are also ready for the challenge of printing houses on Mars. I like their big thinking and fearless attitude. Printing houses on Mars just might catch the attention of a certain space, battery and software entrepreneur that also thinks big and drives a Tesla! Hopefully he is reading my book.

Kabuku Inc. (https://www.kabuku.co.jp/en) prides itself in developing products and services that fuse hardware, software and design with core digital fabrication technology. When you visit their website, use your browser's translation feature, unless you read Japanese. They have an impressive number of 3D tools, platforms and case studies, including a cool 3D printed car they jointly developed with Honda Motor Company - I want one of these as well. In addition, they offer open innovation 3D printing solutions, a 3D print cloud API solution, a 3D scanning solution and a 3D printing partner solution.

They also own Rinkak which is Asia's largest 3D Marketplace featuring a wonderful assortment of 3D printed products for personal and business use.

If that is not enough '3D candy', they offer "Rinkak Business which provides enterprise business critical cloud services and Rinkak Factory 3D Printing Operation Services for Factory Users.

Desktop Metal (desktopmetal.com) is reinventing the way engineering and manufacturing teams produce metal parts, from prototyping through mass production. In 2013, CEO Ric Fulop began collaborating with world-leading experts in materials science, engineering, and 3D printing. Their work together over the course of two years drew multiple independent inventions together to form the basis for Desktop Metal's technology.

Markforged (markforged.com) is also developing affordable 3D metal print systems that can be put into action fast. On their website, they describe scenarios where 90-99% cost savings can be achieved using their printing solution versus the typical manufacturing process.

Autodesk Tinkercad (Tinkercad.com) is an easy-to-use, fun, browser based 3D design tool that allows you to create 3D printable items. I know it is fun, because I enjoy it. It is used by designers, hobbyists, teachers, and children. It can be used to make toys, prototypes, home decor, Minecraft models, jewelry and much more.

The Tinkercad website describes their automated process to move design to printed product: 'When a user wants their 3D model printed, they select 'Order a 3D Print' from the Design menu. Tinkercad then connects to our API and forwards the model to the i.materialise 3D printing interface (called the 3D Print Lab).'

There is no better time for companies and professionals to begin exploring the possibilities of 3D. You can experiment in your office at home or at a Makers facility

Robots and Drones Create Extraordinary Mobile Opportunities

I define robots as internally or externally controlled machines that can perform many types of tasks in the physical world, from very simple to highly complex. They are mobile in varying degrees depending on the requirements of the user. They can learn or they may simply do one repetitive task. Some can think. Robots are our partners and not our enemy.

Fun Fact: The earliest recorded robot was built in 270 BC by an ancient Greek engineer and inventor named Ctesibus. He made organs and water clocks with movable figures.

Robots are no longer relegated to science fiction or the comical 'danger Will Robinson' robot on Lost in Space. Robots are a here-and-now emerging technology that you must have on your organization and career radar. They serve many purposes and are benefiting more and more industries every day. They can reduce expenses, drive revenue and change the way you do business.

The scope of robotics is growing rapidly, including in autonomous vehicles, offices, hospitals and factories. There are even mobile follow-me suitcases and pets!

Atom size machines built with nano-technology are also robotic devices. They can travel through your body and fix organs or alert medical professionals of impending danger. For the most part in this chapter we will concentrate on larger robotic devices.

The goal of this chapter is not to teach you how to build a robot from scratch, although I do briefly describe how you can begin creating simple and useful robots. The chapter is focused on helping you begin to get your arms around the technologies, benefits and possibilities. As always, I urge you to think about how you can leverage the technology in new ways to improve your business, your life and create customer happiness. My greatest hope is that this chapter will provide you with ideas.

In the next part of the book I will demonstrate how robotics are being used in specific verticals such as people movement, commercial delivery and logistics, the smart home, helping disabled people, entertainment, life-and-death crisis scenarios and security. I will include personal successes, specific devices and companies that are on the forefront of robotics.

Robots can be found in thousands of shapes and sizes from an R2D2 in Star Wars to a hotel check-in humanoid or a vacuum cleaner doing the work, so you do not have to. The only limitation is your imagination.

They can have wheels, treads, legs or take on other forms such as a snake (great in disasters and emergency response), a dog or a humanoid-like person. They can nimbly travel the halls of an office or school, sit on a desk or provide value on a factory conveyor line. Integrating robots with other types of emerging technologies gives them super powers.

Robots play a central role in the Internet of Things (IOT). Mixing a robot with technologies such as artificial intelligence, big data, computer vision and tactile touch sensors exponentially further increases their abilities. Intelligent robots are rapidly assuming more ambitious roles such as partnering with doctors to perform surgeries and as companions and aids to elderly people that live alone.

Fun Fact: The first robotic surgery was performed using the Puma 560, a robot used in 1985 by Kwoh et alto to perform neurosurgical biopsies with greater precision.

Earlier in the book I spoke of the Maker / Do It Yourself (DIY) movement that is sweeping the world. Makers and school robotics teams are building sophisticated and useful robots from off-the-shelf parts, discarded toys and byproducts of industry. As I have mentioned before, Maker meetings are a fantastic way to learn about technology, increase career skills, build friendships and have fun.

I have attended numerous STEM, high school and collegiate robotic competitions and demonstrations. The high level of functionality of the robots is impressive. These are the future leaders of the robotics industry. If you have a local competition I urge you to attend or sponsor the event.

Industrial robots are often comprised of an arm that completes a repetitious task in a factory or warehouse.

Perhaps, welding a part or fastening screws and bolts. They never get tired, need coffee or a vacation day.

Mobile industrial robots can traverse warehouses and factories delivering parts and picking and packing even the heaviest products. If they get hurt or break down it is far less of a concern than a person getting hurt. Safety first! Simply replace the part and robots are ready to get back to work.

Amazon uses a fleet of 100,000+ small orange robots originally developed by Kiva in their warehouses. The micro-bots automatically pick packages and bring them back to people that pack them in boxes. This saves a great deal of time and money. It also reduces errors. Amazon quickly realized return on investment from the little orange robots and how bright the future is for robotics, so they bought Kiva for $775 million.

Imagine how you can improve security and safety in your organization by doubling or tripling your ability to view threats without adding to your employee roster. Robots can be programmed to patrol hallways, offices, warehouses and factories. They will work 24x7x365 and never complain.

Security robots armed with cameras and mix-and-match sensors take situational awareness to a very high level. One or many mobilized robots can stream video, pictures and many types of important metrics such as heat and tremors to a central command center anywhere in the world. They can intelligently climb over and go around objects in their path.

Soft snake robots with cameras can slither into pipes, drains and crevices too small for people and larger robots. This can be critical during a crisis event. Nature lends a hand in providing innovative ideas for designers creating all sorts of new products.

Use cases for these security / crisis management robots include break-ins, active shooters, bomb threats, earthquakes, tornadoes and many more disruptive events. People safety is always our number one concern and anything we can do to protect them is worth the effort. Security robots are often military grade, tough and durable. They can take a pounding and happily do their jobs. Prices for security and industrial robots are decreasing dramatically.

All the digital security benefits I described above can be monetized. When you are safe and resilient it increases your ability to fend off threats and increases the continuity of your business. This is a differentiator when compared to less secure organizations. It should be incorporated as part of your pitch to prospects. It can generate additional revenue.

Historically, large industrial robots were isolated from humans. More recently, collaborative robots are being developed that can work side-by-side with humans in industrial and office settings.

For example, in 2012 Rethink Robotics (www.rethinkrobotics.com) released Baxter, which allows people to work side-by-side with robots. Baxter is powered by simple to use Intera software. It can be programmed by the robot owner instead of by expensive robotic experts. Baxter uses computer vision to accomplish complex or simple tasks. Baxter is smart. Training him is easy and fast. Baxter even has a friendly smiling face and never gets upset. On the Rethink web site, you can meet Baxter's friends that do many valuable things including Sawyer and the 'gripping' Clicksmart family.

Office robots can be a physical extension of an employee working from home or half-way around the world. For example, Double Robotics (www.doublerobotics.com) has a line of telepresence robots that enable you to participate in the office or classroom school physically being there. It takes virtual work to a much higher and more interactive level.

These telepresence robots include video screens and cameras. It feels as though you are in the room with people located anywhere in the world. You can even participate in meetings, classes and coffee breaks through your robotic telepresence. There are some interesting case studies of companies, such as LinkedIn, that effectively use telepresence robots.

Office robots can be lifelike with realistic faces and synthetic skin. They can scoot down the hall to deliver mail or clean the trash.

Robots can work 24x7 and will not complain if they do not get holidays off. This can be advantageous as the United States starts on-shoring work and jobs that were previously off-shored. Automation using robotics will increase revenue and productivity and reduce costs. No matter how little people are unfairly paid overseas, machines can do it cheaper, faster and better.

Robots in the home can be saviors to the elderly, disabled or anyone who needs help. For the elderly, a smart home may be an alternative to living in a nursing home. We will discuss that value as well as other home IoT devices in the upcoming chapter, 'Smart Home and Why It Matters'.

Home robots can also help us in smaller ways. They can bring us drinks and snacks while we are watching TV. Imagine asking your robot to get you a soda or a beer.

I use a robotic dog feeder when I go away for the day. Anyone that owns a dog can relate to the feeder as freedom.

Suitcase are now smart. ForwardX CX-1 (Forwardx.com) is a self-driving suitcase that was demonstrated at the 2018 Computer Electronics Show (CES). Wheels on a suitcase used to be cool but a self-driving suitcase is super cool! It will follow you at an airport, train station or anywhere else. It uses computer vision and RF technology to auto-follow. Amazingly, it has 4-wheel drive, which is better than my car! It makes use of Internet of Things (IOT) sensors to avoid obstacles. It comes with a battery that you can also use to charge your devices.

Recently, my son bought a robotic Segway miniPRO (http://www.segwayminipro.com/), which is moderately priced, especially for the functionality it delivers. Segway calls it 'The Hands-Free, Self-Balancing Personal Transporter'. I am not endorsing it or suggesting you go out and buy one unless you do your due diligence.

It has a base that is mobile and stands upright using gyroscopes. It does not need a handle for balance. Compared to a standard Segway that you might see security guards riding in malls, the miniPRO is smaller, lighter and easier to use.

It can also be controlled with an app without physically being on the device. Hmmm, I took one look at it and thought, this can be modified to be a pretty cool business robot. I checked the Segway site and sure enough there is a tricked-out version being sold as a robot. I can see some innovative entrepreneurs customizing it for specific niche markets and selling it at a premium.

My son and I recently hit the streets. He rode the Segway miniPRO and I went on a training run for an upcoming race. He attached his mobile phone to the Segway and filmed me from 360 degree angles. I had never watched myself run, except for brief finish line videos at the end of races. The video was transforming and enabled me to make immediate improvements to my technique which resulted in a personal best time in my next 5k!

iRobot (http://www.irobot.com/) is famous for the Roomba, which has made life easier for millions of us while keeping our floors clean. They are a world leader in robots and besides the Roomba, they make robotic mops, pool cleaners and military robots. They are an innovative organization committed to invention, discovery and applying technology for practical uses.

Google purchased Boston Dynamics (https://www.bostondynamics.com/) which vaulted them to a robotic leadership position. In addition, through their Waymo technology they are leading the race to a truly autonomous car. There is much more information on autonomous vehicles in the chapter, 'Smart Personal Transportation', in the next section of the book.

Ozobot (https://ozobot.com) makes pocket size robots that teach people about programming robots in an interactive manner. They are popular with educators and in STEM programs. I feel these small robots are valuable to anyone that wants to learn and increase their skills by digitally playing.

Ozobot is the type of gift I suggest you consider buying for young people. Learning about tech, programming and robotics with a tool like this can be the spark that creates a love for technology, innovation and possibly a very rewarding career.

I will share additional interesting real-world uses for robots as part of integrated smart solutions in the next section of the book.

Drones

A drone is a robotic unmanned aerial or underwater vehicle. Drones come in all sizes from less than an inch to the size of a full-scale plane. They derive intelligence using AI, cameras, sensors and even virtual reality!

Drones will create more than 100,000 jobs and at least $482 million in tax revenue for the United States by 2025. For the drone industry to prosper, safety must be a top concern and regulations protecting people and property must be considered. Congressional hearings in the United States have already occurred on these issues.

Drones are becoming a key tool for security and safety. I recently met with a startup company that automates the flight path of security drones. No human ground piloting is required. The drones are programmed through a friendly interface to fly predefined routes.

A fleet of such drones can survey multiple locations simultaneously anywhere in the world and send video streams that are consolidated in a central command center console. For example, multi-national organizations can have drones in Israel, India and Indiana fly routes that stream valuable real-time images and other data to their emergency command center in Idaho or Iceland.

I would be remiss if I did not mention that, as with any tool or technology, some people will look for nefarious ways to use drones. Drones have already been used for terror related events and this will increase. Drone delivery can mean delivering packages you buy or bombs. Large drones can carry heavy payloads and drones are much easier to purchase than getting a pilot's license.

Agencies such as Homeland Security must have these types of threats on their radar. There are already devices that can automatically bring down drones that fly in a secure area, but in many cases these devices cannot be used, as it would constitute air piracy or breaking other regulations.

Returning to the many benefits of drones, they are already delivering medicine and other critical supplies to areas impacted by natural events, such as earthquakes or tornadoes.

Drones can be used for border control and to monitor large sporting and entertainment events. Consideration for privacy will be an area of concern that must be resolved.

Imagine the value in autonomously surveying thousands of miles of oil and gas pipelines to thwart threats or mesh drone data with artificial intelligence to predict maintenance opportunities before a breakdown.

Off-the-shelf commercial drones costing from $99 to $2,995 can provide great value for the right organizations. These drones typically include cameras and even virtual reality interfaces so you are seemingly in the pilot seat. You could even hack these drones to add additional sensors for custom capabilities that can benefit your organization or customers.

Hubsan (http://www.hubsan.com/na/) is a leader in cost effective drones (many under $200) with an impressive array of features including app compatibility, cameras, GPS, flip-and-roll, automatic take-off and automatic return home.

BFD Systems (https://www.bfdsystems.com) works with businesses on some serious custom drones. These include agriculture, security and delivery type drones. They have many pictures and cool stories on their web site.

CyPhy Works (https://www.cyphyworks.com) has an interesting niche as a leader in tethered persistent drones that are attached to a wire. They cater to industries such as defense, public safety & security, oil & gas, telecommunications, mining & construction and the media. You might think, as I did, that being tethered has disadvantages but it enables their PARC drones to fly for hours instead of minutes and provides a great deal of value in many ways.

For example, for the second year in 2018, CyPhy Works provided persistent aerial views of the Boston Marathon (there is a video on their web site). The cold weather, rain and winds made it difficult for racers, law enforcement, and marathon officials. The conditions kept most air assets grounded but CyPhy drones flew for hours. Their live stream video into the Emergency Operations Centers (EOC) provided hundreds of local, state, and federal agencies with the ability to detect, assess, and respond to activity along the route.

Their solutions are portable, weatherized, and can be deployed quickly, making it ideal for 24/7 surveillance and situational awareness. This is vital for marathons, outdoor concerts, sporting events and any large gathering for safety and security. I even had an idea for a tethered drone to monitor local traffic patterns, in cases where it is not cost effective or practical to use helicopters.

Drones are transforming farming. Harper Adams University in the United Kingdom attempted a project called Hands Free Hectare in 2017 (http://www.handsfreehectare.com), which aimed to be the first in the world to plant, tend and harvest a crop with only autonomous vehicles and drones. It ended with a successful harvest. No humans were used in the process. In 2018 they are attempting to improve on the first harvest with more precision, while harvesting a hectare (2.471 acres) of wheat.

In autonomous precision farming, small driver-less tractors plant crops and drones integrated with cameras and sensors to pinpoint land that needs water or seed as well as crops that are ready to be harvested. Robotic pickers can automatically test for ripeness and pick fruits and vegetables at the optimal time. Precision farming replaces the shotgun approach where water and other liquids were sprayed even where they were not required. Using a series of smaller than the old way tractors, saves money, creates resilience and is more sustainable. This smart new precision approach replaces waste, lowers expenses and generates revenue. It is another case of emerging technology and automation doing it faster, better and cheaper.

Commercial drone deliveries are here and will be more common soon. Where it adds value, they will partner with autonomous trucks to provide door-to-door autonomous deliveries. In the 'Smart Supply Chain' chapter I describe how companies such as Amazon are already beginning to make this a reality.

Drones can also entertain us. Hopefully, you were fortunately to watch the illuminated drone swarms that stole the show during the 2018 South Korea Winter Olympics opening ceremony.

Drone racing is becoming a popular sport. I can see why as I always appreciated watching radio controlled (RC) airplanes and even owned some simple ones. Watching drones in flight is very cool for spectators. For the pilots using augmented reality to virtually put them in the pilot seat it must be amazing as their drones go through gate-type obstacles at high speeds. In New South Wales, Australia and Las Vegas, NV there are indoor drone arenas. Many more outdoor and indoor races will be popping up.

I used a large drone to film a video for a corporate business program kickoff meeting. The film wowed 40 executive attendees as it flew over our locations revealing some amazing overhead views. I received many questions on how they could use drones for marketing and sales.

I have fun tiny lightweight drones through my house for fun. I have a 19-foot ceiling in my living room which is lousy when it comes to changing the smoke alarm, but is very cool for indoor drone flights. I am also researching follow-me-drones that can accompany me on my training runs and perhaps during 5k races. I think other runners would love to see aerial views. Perhaps there is a side-business there? For a company already organizing races this can be an additional revenue stream. I already pay to get a high-quality finish line picture in races and would pay a lot more for an in-race follow-me video.

NASA, a leader in drone development, selected three aeronautic teams to explore ambitious drone ideas. "Our idea is to invest a very modest amount of time and money into new technologies that are ambitious and potentially transformative," said Richard Barhydt, NASA's acting director of the Transformative Aeronautics Concepts Program (TACP). "They may or may not work, but we won't know unless we try."
The studies will explore whether and how it might be possible to:
1. Build a path toward safe inclusion and certification of autonomous systems in aviation. Autonomous systems, such as self-driving cars and future UAS, rely on learning algorithms that adapt to new goals and environments. The idea

is to develop autonomy-enabling algorithms that lay a foundation for establishing justifiable confidence in machine decisions and, ultimately, lead to certification of autonomous systems.
2. Develop new methods and technologies for a remotely-piloted drone to make sure it's "fit to fly" before every single flight. The idea is to verify the aircraft is structurally and mechanically sound, and that all its onboard systems have not been damaged or hacked in some way. If it's not sound, the aircraft will ground itself.
3. Use quantum computing and communication technology to build a secure and jam-free network capable of accommodating hundreds of thousands of drones flying each day. Because of the way data is organized and processed, quantum computing enables certain computations and communications to be done much more efficiently than a regular computer. For example, quantum computers may be able to solve certain problems in a few days that would take millions of years on the average computer.

Drones are not limited to the sky. **iBubble** (htttps://ibubble.camera/) makes sleek looking underwater diving follow-me drones that their site mentions will begin shipping in late 2018. These are intended for scuba divers, free divers, professionals, hotel resorts and anyone that can creatively think about using underwater drones. They are not intended for the backyard above ground pool. I like the niche they are carving. There are endless uses for these devices.

Now is the time to start thinking about how you can incorporate robots and drones for your business or personal enjoyment. We are only scratching the surface of possibilities for these technologies.

Blockchain
Transforming an Ancient Concept

The practical consequence is for the first time, a way for one Internet user to transfer a unique piece of digital property to another Internet user, such that the transfer is guaranteed to be safe and secure, everyone knows that the transfer has taken place, and nobody can challenge the legitimacy of the transfer. The consequences of this breakthrough are hard to overstate.
Marc Andreessen, the co-creator of Mosaic which was the first popular web browser

The value of Blockchain is way more than that of Bitcoin!
Marty Fox

Blockchain technologies have the power to disrupt many industries. Now is the time to begin investigating how you can use it. Many people are unsure of what blockchain is. They think it is another word for Bitcoin. This lack of awareness can be benefit you.

As you read through this chapter allow your imagination to freely explore how you can apply blockchain to your industry. This chapter leverages my research and insights and information from The National Institute of Science and Technology.

Simple Definition:

Blockchain offers the first digital spin on age-old paper ledgers used by banks and other institutions for thousands of years.

It is a decentralized ledger that maintains transaction records on computers that make up the chain. All the computers are in sync which produces many copies of the chain. When a group (block) of records is entered in the ledger, the block's information is connected mathematically to other blocks, forming a chain of records.

The beauty of the blockchain is that the information in a block in the chain cannot be altered without creating a discrepancy that other digital record-keepers in the network would immediately notice. Because of these checks and balances, blockchain technology produces an immutable (cannot be modified) ledger record without record-keepers having to know or trust one another. This transparency eliminates the risks inherent in data being stored in a central location by a single owner. In the real-world it eliminates, for the first time, the need for a middle-person, such as a bank.

If you are in the financial, real estate, legal or various other professions, blockchain may be in your future, which is why it must be on your radar.

Why it is important:
Although Blockchain gained popularity due to its close ties with Bitcoin it will have far broader impacts on our global economy. Use-cases in this chapter illustrate how blockchain, even in these early days, goes far beyond being the platform for cryptocurrencies.

There is a great deal of hype around blockchain. Some companies are changing their names to include blockchain to create excitement. When you are exploring the market for potential blockchain partners/vendors/investments make sure they have experience and are not simply jumping on the 'blockchain bandwagon' for a quick score.

Blockchain is a very complex set of technologies for developers and there are challenges. There will be many career opportunities for developers that can program blockchains.

Some companies are getting into the blockchain game to 'mine' coins. Mining coins is probably not what you think it is. It sounds cool and gives the image of some dusty old prospector using a pan during the California gold rush and perhaps striking it rich. In reality blockchain mining consists of many computers working together to solve complex mathematical equations so immutable ledger entries can be completed. It used to be easy to mine coins with one of two computers. Now it takes many computers and a lot of electricity to profitably mine them.

Blockchain Deep Dive (Warning a little bit of lower level 'tech talk' below, but I promise to keep it to a minimum):
At its most basic level, a blockchain enables a community of users to record transactions in a ledger that is public to that community, such that no transaction can be changed (there is that immutable term again) once published. If an entry is recorded and needs to be changed,

the original record is not modified, rather a new record is created with the new data.

In 2008, the first blockchain was built using several technologies and computing concepts to enable the creation of Bitcoin, the original cryptocurrency.

Cryptocurrency is electronic money protected through cryptographic mechanisms. Blockchain currency systems are novel in that they store value, not just information. The value is attached to a digital wallet—an electronic device (or software) that allows an individual to make electronic transactions. The wallets are used to sign transactions sent from one wallet to another, recording the transferred value publicly, allowing all participants of the network to independently verify the validity of the transactions. Each participant can keep a full record of all transactions, making the network resilient to attempts to alter that record (or forge transactions) later.

Architecture Pluses and Minuses

I am a database architect and I generally preach the value of centralized data in organizations to eliminate silos and allow organization to extract the most value from their data. My feeling is that the blockchain, being a distributed system, has pluses and minuses as far as an architecture. On the plus side as a platform there is a lot of transparency which fosters trust. Currently there is quite a bit of mistrust to overcome. It also has a high degree of security built-in

On the minus side, blockchain technology currently (2018) has limits on the amount of data that can be stored and is not meant to be a general storage medium. To quickly calculate complex hashes (cryptography and mining) on transactions and distribute transactions amongst the network, transactions need to be relatively small.

Large amounts of data are usually stored "off chain," with "pointers/references" or hashes of the data stored within the blockchain itself. Blockchains also benefit from data being immutable, which is not a trait general purpose data usually needs. Immutable means it cannot be changed. In some use cases that is important but in others it can cause problems.

To create complex security hash codes and keep the blockchain in synch on hundreds or thousands of computers in near real-time, which is critical, takes a tremendous amount of computing power and electrical energy. One estimate is that implementations of blockchain utilizes more electricity than the country of Singapore. Some blockchain mining companies are locating their data centers near cheap hydro-electric sources of electricity to keep expenses as low as possible. A medium size city in upstate New York disallowed blockchain mining as companies were relocating there and driving up the cost of electricity for residents. As blockchain use explodes that electricity demand may increase exponentially.

Additionally, as of this writing, there is latency in recording transactions to the blockchain. For uses that demand real-time transactions that is not tolerable. There is research being done to significantly reduce latency to near real-time. For example, graphene block technology may offer significant improvements compared to existing blockchain technology in size and speed.

Next Steps:

If your organization is thinking about implementing blockchain technology you should do a lot of research to understand important aspects of the technology. For example, what happens if you want to make modifications to the data stored? In a typical centralized database, this can be accomplished easily but organizations need to understand the difficulty in changing anything that is already on the blockchain.

Another critical aspect of blockchain technology is how the participants agree that a transaction is valid. This is called "reaching consensus" and there are many models for doing so, each with positives and negatives for a specific business case. All of this will be figured out in due time.

Some existing blockchain technologies focus on storing wealth, while others are a platform for smart contracts. New blockchain technologies are being developed constantly to enable new use cases and to improve the efficiency of existing systems.

Some blockchain implementations are permission-less, meaning anyone can read and write to them. Other implementations limit participation to specific people or companies, allow finer-grained controls and may be managed by a central entity. Knowing these specifics allows your organization to understand what will be most applicable to your needs.

Use Cases and Early Adopters

Below are a variety of blockchain use cases I have compiled from public and my private sources. Keep in mind this is only a small sampling of the possible uses of the blockchain. Some are in the early stage and may or may not succeed. I am sure there will be many interesting new blockchains created in the coming months and years as the technology matures.

As you read these think about how your organization may benefit from using a blockchain OR creating a blockchain with partners in your niche or a different niche. If you are a technologist you may want to learn blockchain skills that will be in demand as the technology matures.

Trusted Timestamping

This is a reliable way to prove that information existed at a given point in time. The use of a blockchain allows a party to prove they had access to a piece of data in a way that cannot be repudiated. For example, if a person had to prove they had possession of a file, a task was completed on a certain date, a contract was signed, or proving any other event occurred they could record a hash value as an annotation to a transaction. Then, if he or she ever needs to prove possession of the file, has been recorded publicly. This may be useful for the legal and real estate professions.

Banking

Blockchain technology will empower banks to adapt or even completely change their practices to focus on being platforms for value exchange and not just places to store value.

Here is interesting example of modifying the characteristics of a traditional blockchain for the benefit of the participants:

Suppose several banks or credit unions want to keep a private, distributed ledger available to only the participating banks. This would provide the ability to record transactions from each bank in a way that is visible to the participants, but not to the public. However, to do this as a private blockchain (to avoid having to use an expensive proof of work algorithm), each bank takes turns signing the blocks under a distributed consensus algorithm.

In addition, the private blockchain members could over-ride the immutable record feature if it was beneficial to their use case.

Supply Chain

A lot can go wrong on the product journey. As a director in the logistics, technology and business continuity worlds, I see great value in the future use of blockchains applied to the supply chain.

Recording the transfer of physical goods as they move through the supply chain from a producer, to a shipping terminal, to a ship, to a cargo train, to a delivery truck to a warehouse, to inventory or to deliver to a store is an important application and benefit of blockchain technology.

A blockchain could play a crucial role in trust and transparency through the entire supply chain from suppliers to downstream customers. The blockchain could be used to monitor supplier actions. Suppliers can record the product produced on a certain day in a way that other viewers of the blockchain can verify. With a blockchain, it is possible for warehouses to manage inventory and logistics efficiently to avoid overstocking. The transparency and verification can extend to the buyer accepting the product.

IBM produced a cool commercial that shows a tomato moving through the supply chain from farm to your kitchen. You can probably find it on YouTube.

Walmart is working on a blockchain that will monitor the movement of pork within China that will help verify contracts at each step of the supply chain.

Provenance is working on a blockchain to track seafood every step of the way from where it is caught to the buyer. This will insure the product came from the intended source and not a poacher and was shipped properly.

Blockchain in Transport Alliance (BiTA) - (https://bita.studio/) is a standards organization for the supply chain. Members of BiTA come from every facet of the industry: tech vendors, OEMs, tier-1 suppliers, consultants, banks, carriers, shippers, and brokers.

The BiTA web site describes their mission: 'All companies within BiTA share a unified mission of developing a standards framework, educating the market on blockchain applications, and encouraging the use of said applications through exemplary implementation. In addition to establishing market standards, the Organization is focused on engaging with the industry and providing educational resources to ensure the full potential of the benefits of blockchain technology is carried through. BiTA standards will address smart contracts, freight payments, asset maintenance and ownership history, chain of custody of freight, and other issues facing the industry.

Insurance and Healthcare

Having been an executive with a global insurance company I can attest that whenever someone visits a health care provider, many transactions take place behind the scenes. Administrative transactions from nurses, doctors, staff, medical providers, insurance companies, and pharmacies could all be written to a blockchain. Transactions, such as checking benefits, eligibility, coverage, and the available medicine supply could be logged and read from the blockchain.

Currently, records of these types of transactions reside in disparate systems, sharing results at the end of an often manual process. There are many destructive issues with valuable data (gold) being maintained in disparate systems that do not 'talk to each other'. Doing so can miss creating customer value. Instead it often leads to frustration, errors and needless expense.

Civic (civic.com) has an identity blockchain that allows consumers to authorize the use of their identities in real time. It is designed to facilitate on-demand, secure, low-cost access to identity-verification services via the blockchain. Civic partnered with MedCredits to verify the identity of physicians who want to be listed in their public physician directory.

Energy Industry

Recording autonomous, machine-to-machine transactions regarding electricity use can be a valuable use of a blockchain. This would allow tracing transactions on a smart grid.

Recording certificates is another blockchain energy use. There are many power plants creating certificates that detail the amount of energy produced for subsequent exchange. As of August 2018, there are problems such as emission certificates being spent twice, as well as the need to address regulatory challenges and provide more uniform access for everybody in the market. A blockchain can track and validate issuance and spending of energy certificates.

Another example is using a blockchain in the trading of excess renewable energy. Buildings can be wired with devices measuring energy usage and recording it to a blockchain. This enables excess energy to be sold and bought on a market. This trading type use-case can be extrapolated to any type of trading platform.

Tracking Donations

The United Nations (U.N.) deployed a blockchain that helps track food and cash for countries in need of assistance. The way it works is, the U.N. sends the money to the proper agencies who send it further downstream to food vendors for the ultimate distribution to people. Each step in the chain is logged in the blockchain. This enables the U.N. to validate that money is being utilized properly and it does not go astray.

Publishing

SolidOpinion.com empowers publishing site visitors to engage with content through article comments and subscription sign-ups in exchange for Engagement Tokens.

Additionally, advertisers use Engagement Tokens to purchase space under relevant article headlines that serve as a natural tie-in to the product or service featured in the ad placement in an auction-based pay-per-article model. The result of this model is deeper relevancy and increased content engagement.

According to SolidOpinions, as of August 2018, clients include the LA Times, Chicago Tribune, Baltimore Sun and others with a base of over 4.4 million users generating over 200M monthly views. The users are already purchasing tokens in exchange for U.S. dollars, generating up to 100,000 transactions every month.

Travel

TravelChain.com is working on a blockchain which provides accessible, authentic and verifiable data to travelers. On their site, they have a detailed business plan with much information on blockchain benefits for travelers. It uses the aforementioned graphene block technology which may offer significant improvements on standard blockchain technology in size and speed.

Data Asset Ownership

Bitmark.com develops tools that define a new digital freedom by providing an economic framework of standardized property rights, rules and infrastructure – just like the ways in which we own assets for the rest of our lives. The system intends to allow individuals to derive value from crypto property just as we do from the things we own in the physical world – selling, buying, transferring, donating, licensing, passing down, protecting, and much more. These tools interface with digital environments where we create and share things: social media, fitness and health apps, productivity and financial software, and much more. The goal is to allow users to simply apply a mark of accepted ownership to a new creation and embed it into Bitmark's standardized, universal crypto property system.

There are many other potential uses and opportunities for blockchain technologies. As you now realize blockchain goes way beyond just powering Bitcoin and digitizing money into cryptocurrency. As the technology matures, blockchain will be used throughout our society. It can be a game-changer.

Augmented and Virtual Reality
Where Anything is Possible

Augmented Reality (AR) and Virtual Reality (VR) are virtual environments that will change the way we interact with everything in our world. The opportunities in AR and VR are limitless and there is great excitement about what will soon be possible in business, training, entertainment, travel, work and every facet of life including:

- Architects can design buildings and homes that prospective buyers can 'walk through' in a realistic environment.
- Retail stores can provide a more realistic and compelling shopping experience that visitors can experience on a deeper emotional level for increased revenue and a reduction of the cost of returned items.
- Astronauts can train by floating through the International Space station and can practice walking on a precise replica of the Mars landscape.
- Students and employees train and learn in safe realistic virtual environments rather than dangerous ones. Imagine training for active shooter response or learning to operate a fork life in a realistic environment. Well, read on and you will see it is available now!
- Movie fans, gamers and tourists can experience immersive excitement often in the comfort of their own home.
- Weekend and professional athletes can 'up their games' in a new and exciting way.
- Historians can travel back in time and futurists can explore the future.

In this chapter, we will discuss what is here now and where these technologies will take us in the near future. Your industry will be transformed by augmented and virtual reality. You must be the organization that leads the transformation. Now is the time to explore how you will leverage these technologies for a competitive advantage.

Augmented Reality (AR)

Augmented Reality (AR) is a type of Virtual Reality (VR) that allows virtual images to be superimposed over the real-world environment, providing users with nearly unlimited experiences. Augmented Reality is sometimes referred to as real-world or world sense reality.

AR can be used to try out new furniture in your home before purchasing or for surgeons to float x-rays in the air so that they can be easily viewed while performing surgeries.

Business users can project content and images into a room or onto a wall for a realistic interactive experience. Office applications can include floating resizable spreadsheets and documents.

I sometimes refer to AR as BR (Better Reality). The ability to benefit from enhanced information against a real-world backdrop is valuable. You might be thinking, well Google tried it with Google Glass and it failed. As you know from the mindset I posed earlier in the book, smart 'failure' is to be coveted. Google Glass was an experiment, perhaps even an MVP (minimal viable product). It will come back in an improved form and it will be a winner.

While fully immersive virtual reality (VR) entails creating the entire world, augmented reality enhances the world around us. It is simpler and faster to build realistic environments. Sometimes it is referred to as mixed technology.

Pokémon Go is an example of a simple 2D augmented reality. It uses geo-spacial real-world location coordinates to place Pokémon characters in public places in the real world, perhaps in a shopping center or park. Pokémon players only need their smartphone to view and try to capture the characters. No headgear or computers are required. Players get points for every Pokémon they capture. In 2017 Pokémon went viral and built a user base in the millions.

In 2018, while researching this book I installed the Pokémon app and played it. It was fun but I was awful. I first tried it in the supermarket and shoppers must have thought I was either drunk, crazy or both. I stumbled left and right, knocked down a stack of cereal boxes and had a lot of folks rolling their eyes. But my research was a success! I caught a Pikachu and had a story to tell you!

If you watch sports on TV you are already very familiar with augmented reality, although you may not realize it. The yellow stripe superimposed on the field that indicates how far a team has to go to get a first down is created with augmented reality. It has been used for decades. I do not know about you, but on plays when it is not displayed I miss it! I think it would be so valuable if players could see it in their helmets, if that is legal. Perhaps a near future Google Glass type product will make that a reality.

Another AR example in sports is the virtual cube superimposed over home plate while watching a baseball game. It is visible to TV viewers but not to the fans in the stadium. It provides a precise indication of balls and strikes and can sometimes be embarrassing when umpires 'miss one' and call it incorrectly. Eventually this type of technology will help disrupt the umpire and referee professions.

Have you ever watched a baseball game on TV and there is an advertisement on the wall to the left or right of home plate that changes from inning to inning? Well, that is another use of augmented reality. For advertisers, it is great to have the option of only buying adds for a few innings instead of having to commit to buying one expensive ad for the entire game. Fans benefit by receiving multiple offerings throughout the game.

During the 2018 Winter Olympics, I enjoyed augmented reality while watching the giant slalom. The TV network superimposed colorful visual paths skiers took down the mountain to point out how some went wide and other went down the middle. I am not a huge skiing fan but I could not turn away and watched 20+ skiers zoom down the slopes. It was the augmented reality that lit me up.

For people that cannot physically travel or prefer not to, augmented reality offers the opportunity to explore distant lands without leaving the comfort of their smart homes. Many cities are creating detailed augmented landscapes and museums, which can be enjoyed from anywhere. Augmented tourism will become a big business in the near future.

Augmented reality also allows you to take realistic trips down memory lane. Imagine touring your old neighborhood or city with augmented informational pop-ups describing the landmarks that were there in the past.

For example, I recently met with a tech company in lower Manhattan, near the New York Stock Exchange. I spent a great deal of my early career in lower Manhattan working in technology for major financial institutions. I had not been down there in many years. I walked the streets which brought back so many wonderful memories of people and places long forgotten. It would have been so cool if I could have enjoyed an augmented experience of sightseeing buildings and viewing them as they were at any year I indicated, instead of as they are now. Imagine doing that with any sightseeing trip to any city. It will happen and sooner than we think.

Another potential use for augmented reality is Big Data analysis and visualization. The benefits of visualizing data in real time on a life-like scale, and being able to interact with the data (immersive visualization), are endless. Organizations that create and analyze large amounts of data are limited in their analysis and data interactions. Visual angles (i.e. only being able to view data on a flat 2D surface,) navigation capabilities, and human perceptions are limited in a small 2D environment. AR immersive visualization solves these problems. The data is projected in 3D and the user can step inside of the data sets and view, edit, manipulate, and analyze it in a dynamic, real-world environment.

As we discussed in the drone chapter, professional and hobbyist pilots can use AR to virtually travel in the cockpit of their drone as they operate it remotely. This feature is not limited to expensive drones, some models under $150 offer it. Wow, imagine the adrenalin rush of flying through racing gates or over lakes and mountains rather than watching from the ground.

Augment (augment.com) is a company that brings a physical presence to online shopping. They enable you to display products across e-commerce platforms in life-like, 3D augmented reality to drive sales and engagement. On their site, they have customer stories in manufacturing, retailing and consumer packaging. They also offer an application programming interface (API), software developers kit (SDK) and guides to help you develop your own solutions around their engine.

I also downloaded their free app and quickly created my own simple AR. I superimposed giant butterflies sitting on a railing and tigers roaming through a Long Island Railroad station platform and sent it to my wife and son. You can take it to much more advanced levels, but it pays to start with a few quick and fun wins. Thanks to my tech savvy friend Harvey for mentioning this cool app to me.

Microsoft HoloLens has the potential to take AR to a lofty level. It allows you to build, float, place and interact with 3D holograms anywhere in your real environment and from any angle! If Pokémon is on the simple, yet incredibly popular and lucrative end of the AR spectrum, then HoloLens is at the very edge of advanced AR. For example, there is already development by third parties to leverage HoloLens for spinal surgery.

All the necessary AR components are self-contained in the HoloLens headset. It is like wearing a supercomputer. You are not tethered to a desktop, you are free to use HoloLens to explore any type of distant or enhanced environment in realistic detail. HoloLens does not rely on keyboard input, rather it uses more intuitive gesture controls and spacial sound.

The plan for HoloLens is to use a custom Holographic Processing Unit (HPU) that will include an AI processor enabling deep neural network benefits. This powerful benefit will empower HoloLens to visually analyze and recognize objects and do predictive analysis locally, rather than sending data to the cloud and experiencing latency. Immersive AR and VR experiences depend on speed so anything that can be done locally is advantageous. It also reduces risk by not depending on the cloud to be accessible in a crisis event, just when you might need HoloLens the most.

NASA (I mention them often in the book, as they do so many exciting things) is using HoloLens to create a realistic Mars environment here on earth in preparation for future flights. Astronauts can interact and travel through an immersive virtual representation of the Mars landscape, as if they were there. Also, HoloLens will take anyone that wants to go to Mars on the trip as part of their Mars 2020 initiative - without leaving home!

As of August 2018, HoloLens is expensive at $3,999. It is a developer and business tool at that price. I believe the price will drop significantly in 2019. Businesses that are ahead of the curve in using the tool will benefit. Programmers that can build HoloLens experiences will be in high demand.

HoloLens has a potential competitor in Magic Leap. They are a small and well financed company in Florida with a knowledgeable management team.

As of August 2018, Microsoft and Magic Leap have work to do to make the experience fully immersible but the promise is there and if either company or both can pull it off, it will change the way we interact with the world.

Virtual Reality (VR)

Virtual Reality(VR) is the use of computer technology to create a simulated environment. Instead of viewing a screen in front of them, users are immersed and able to interact with 3D worlds. VR is currently used in entertainment, art and design, gaming, education, and tourism fields. It is certain to continue to evolve into many aspects of our daily lives.

Virtual Reality (VR) differs from Augmented Reality (AR) as users are immersed in a virtual world, rather than augmenting the real world. VR will be used in fully immersive movies and video games. You will be part of the story and you will experience the virtual environment first-hand. If you look right of left or turn around - the virtual world seamlessly reacts in real-time. Movie and game production companies are already experimenting with the technology. If you currently enjoy watching movies in 3D or IMAX, soon virtual reality will amp up the fun and realism to a much higher level.

The 2018 blockbuster movie, 'Ready Player One', which I very much enjoyed, is a story about interaction, competition and intrigue in a fully immersive VR universe called the OASIS. It takes place in 2045 in a downtrodden and dystopian physical world on the brink of chaos. OASIS is the salvation of the world's inhabitants. It networks people in a realistic virtual world of excitement and vivid colors.

I believe all the technology in the movie will be available before 2045. Although it seems impossible there is nothing that cannot be created in by then or earlier. If you happen to see the movie, also check out the cool real-world (not virtual) drone pizza delivery in the first 5 minutes. From then on I was hooked!

VR is a perfect example of a technology made possible by Moore's Law. In the past, chips were not fast enough to create a smooth transition as participants redirected their view. That would result in latency, ruining the immersive experience and sometimes causing dizziness and nausea. The exponential advancement in chip processing speed now makes VR possible.

VR headsets do not have to be expensive. Google cardboard is very popular and inexpensive. It works with many mobile apps and creates an immersive experience. There are many makers of low-cost plastic headsets as well. These types of devices can be an entry point to begin getting familiar with VR.

On a much higher and more immersive end of the virtual hardware spectrum we find Oculus Rift. It is a leader in VR technology and a great example of how successful a resourceful individual innovator can become.

Palmer Luckey was interested in electronics and engineering from a young age. Beginning at the age of 14 he took community college courses at Golden West College and Long Beach City College. He experimented with electronics projects including lasers and Tesla coils in his early teens. By the way, this level of passion is not unusual in young people. I have experienced it first-hand and it is a wonderful experience to be part of. I suggest encouraging young people to follow their passion and help them achieve success in any way you can.

Palmer had great interest in virtual reality. He built over 50 different head-mounted displays. He was ambitious and to fund his experiments he fixed and resold damaged iPhones and worked as a computer repair technician. While attending Cal State University as an undergraduate, he worked as a part-time engineer at the Mixed Reality Lab in the Institute for

Creative Technologies (ICT) at the University of Southern California as part of a design team for cost-effective virtual reality.

He was frustrated with the existing head-mounted displays being sold. They had inadequate low contrast, high latency, low field-of-view, high cost, and extreme bulk and weight. He had a vision of what could be and started experimenting with his own designs. At age 17 he finalized his PR1, his first prototype in true inventor fashion in his parents garage in 2010. It featured a 90-degree field of view, low latency, and built-in haptic-touch feedback. How cool is that!

Palmer continued to develop his knowledge, skill and the usefulness of his VR prototypes. His 6th-generation unit was named the "Rift" and was intended to be sold as a do-it-yourself kit on Kickstarter. He formed Oculus VR to facilitate the launch of the Kickstarter campaign.

John Carmack of id Software, who developed the Quake videogame series, asked for and received a prototype headset. He used it to demonstrate a modified version of id Software's Doom 3 BFG Edition on the device at the Electronic Entertainment Expo 2012. The aftermath of this brought great attention to Palmer and Oculus Rift.

The Kickstarter campaign raised an impressive US$2.4 million, almost 10 times its original goal. Oculus VR expanded, taking on more employees and a larger office space. Luckey continued to improve the product."

When Mark Zuckerberg took great interest in the future of virtual reality he envisioned Oculus Rift as the product that could help him meet his goals on his social media platform and broader entertainment arenas. Facebook acquired Oculus VR in March 2014 for US$2 billion. Palmer, 21 at the time, continued to work at Oculus VR until March 2017.

Autodesk (https://www.autodesk.com/solutions/virtual-reality) a long-time software tool developer for designers will be a leader in the virtual space.

They are already reshaping the world of design in AR and VR. Using their current tools, you can transform 2D designs into interactive, immersive digital models, giving context to your digital information.

Their tools enabled engineers, designers and builders to quickly and easily turn CAD data into interactive, real-time

AR and VR experiences. They have a library of information and examples on their web site. These types of tools enable you to create a better and more emotional experience for people using or deciding to buy your products.

eon reality (https://www.eonreality.com/platform/eon-creator-avr/) - The EON Creator AVR Enterprise and Education content builder empowers non-technical users to create compelling AR and VR applications quickly in minutes, not weeks. You do not need to program to create. It empowers people to accelerate learning and improve performance, safety, and efficiency.

Begin experimenting with AR and VR. It is the best way to understand the capabilities and the user experience. You may be in an industry where it is not clear how you can use the technology until you play with it and the light bulb goes on.

Quantum Computing Hardware Leaps Forward

"It's a game changer for the corporation, it's a game changer for our customers, and ultimately it's a game changer for humanity."
Greg Tallant, Research Engineering Manager, Lockheed Martin

"We actually think quantum machine learning may provide the most creative problem solving process under the known laws of physics."
Hartmut Neven, Director of Engineering, Google

This chapter will give you a good flavor for the vast benefits quantum computing will deliver. If the information peaks your interest, you can easily dig deeper by visiting the companies mentioned below.

Ultimately, quantum computers will be millions of times faster than the fastest supercomputers computers now in production. A quantum computer from Google is already 100,000,000 times faster than a high-level laptop computer. They will enable powerful applications far beyond what is currently possible. A simulated human brain, extremely accurate weather forecasting and robots-with-feelings just scratch the surface of what will be possible.

Argonne National Labs describes quantum as follows: Classical computers are number crunching machines, performing basic arithmetical operations on numbers. In computer language, these numbers are expressed in binary number units of zeros and ones, also called bits. Each bit, stores the smallest piece of information and can accept a value of either 1 or 0. Quantum computers are designed to operate on quantum bits. An extraordinary property of qubits is that they can be of any value equal to or between -1 and +1, until we measure them. As in a classical computer, the initial states of qubits need to be prepared before quantum data processing or data storage.

New quantum discoveries and breakthroughs are happening rapidly. For example, Engineers in New South

Wales invented a new architecture based on 'flip-flop qubits', that may greatly reduce the cost of large-scale manufacturing of quantum chips. This will help significantly drive down the cost of quantum computers. You can read about it here: https://phys.org/news/2017-09-flip-flop-qubits-radical-quantum.html

I believe quantum computing will be mainstream by 2028. It surprises many people to learn there are already a growing number of quantum applications now in production. The Volkswagen Group is an example of a forward-thinking company that sees great value in quantum computing. In March 2017, Volkswagen announced its first successful research project completed on a quantum computer: a traffic flow optimization for 10,000 taxis in the Chinese capital of Beijing.

Martin Hofmann, Chief Information Officer of the Volkswagen Group, says: "Quantum computing technology opens up new dimensions and represents the fast-track for future-oriented topics. We at Volkswagen want to be among the first to use quantum computing for corporate processes as soon as this technology is commercially available. Thanks to our cooperation with Google, we have taken a major step towards this goal."

D-Wave Systems (dwavesys.com) is a leader in the field of quantum computing. As of August 2018, D-Wave had installed more than $50 million worth of quantum systems at customer sites. The D-Wave 2000Q system is available through sale or lease as a standalone system, or via their quantum cloud.

Quantum computers are complicated, which is why I like this simple quantum explanation from D-Wave:

"In nature, physical systems tend to evolve toward their lowest energy state: objects slide down hills, hot things cool down, and so on. This behavior also applies to quantum systems. To imagine this, think of a traveler looking for the best solution by finding the lowest valley in the energy landscape that represents the problem.

Classical algorithms seek the lowest valley by placing the traveler at some point in the landscape and allowing that traveler to move based on local variations. While it is generally most efficient to move downhill and avoid climbing hills that are too high, such classical algorithms are prone to leading the traveler into nearby valleys that may not be the global minimum.

Numerous trials are typically required, with many travelers beginning their journeys from different points. In contrast, quantum annealing begins with the traveler simultaneously occupying many coordinates thanks to the quantum phenomenon of superposition. The probability of being at any given coordinate smoothly evolves as annealing progresses, with the probability increasing around the coordinates of deep valleys.

Quantum tunneling allows the traveler to pass through hills—rather than be forced to climb them—reducing the chance of becoming trapped in valleys that are not the global minimum. Quantum entanglement further improves the outcome by allowing the traveler to discover correlations between the coordinates that lead to deep valleys."

Rigetti Computing (rigetti.com) is a full stack quantum computing company. They provide quantum developer tools including the Forest Quantum API. They have an active community that brings together researchers and quantum enthusiasts to share, connect and collaborate. If you are want to learn what it takes to develop for quantum as a career or hobby you will find a lot of information and insights.

I suggest you visit their Try Forest page and review the growing platform, tools, ecosystems and solutions being developed. They even have GitHub repositories, documentation and a lot more very exciting quantum goodies! You will not be disappointed:

    1 - Grove is a repository to showcase quantum programs developed using the Forest API.

    2 - Race against a quantum algorithm - Play a simple game that demonstrates a practical hybrid classical/quantum optimization algorithm using Rigetti's computing stack.

IBM (IBM.com) developed a powerful 50 qubit quantum computer in early 2018. The machine will surpass the theoretical power of a classical 'bits and bytes' computer when it reaches a more fully stable stage - possibly it has as you read this. IBM used an innovative quantum process called superconductivity to architect the machine. This allowed for greater stability than previous architectures that manipulated photons.

IBM has a good video on quantum computing plus lots of other information at -https://www.research.ibm.com/ibm-q/

IBM press releases indicate that future applications of quantum computing may include:

- **Drug and Materials Discovery**: Discovery of new medicines and materials by untangling the complexity of molecular and chemical interactions
- **Supply Chain & Logistics:** Optimizing the supply chain by discovering the ultimate routes for deliveries during the holiday season
- **Financial Services:** Isolating global risk factors but finding new ways to model financial data which can lead to better investments
- **Artificial Intelligence:** Making facets of artificial intelligence, such as machine learning much more powerful when data sets can be too big such as searching images or video. Currently searching multimedia is a labor intensive task. Some companies outsource this type of labor-intensive, monotonous and sometimes disgusting work (you can imagine what they are looking to insure is not on their sites).
- **Cloud Security:** Using the laws of quantum physics to enhance private data safety, making cloud computing more secure

Tom Rosamilia, Senior Vice President of IBM Systems, said: "Classical computers are extraordinarily powerful and will continue to advance and underpin everything we do in business and society. But there are many problems that will never be penetrated by a classical computer. To create knowledge from much greater depths of complexity, we need a quantum computer. We envision IBM Q systems working in concert with our portfolio of classical high-performance systems to address problems that are currently unsolvable, but hold tremendous untapped value."

IBM also released an API (Application Program Interface) and SDK (Software Developers Kit) for the IBM Quantum Experience. Informative videos on this page - (https://quantumexperience.ng.bluemix.net/qx/community?channel=videos). This enables developers and programmers to build interfaces between its existing five quantum bit cloud-based quantum computer and classical computers, without needing a deep background in quantum physics.

The IBM Quantum Experience enables anyone to connect to IBM's quantum processor via the IBM Cloud, to run algorithms and experiments, work with the individual quantum bits, and explore tutorials and simulations around what might be possible with quantum computing.

Google is another leader in the quantum computer race to the top. During March 2018, they released a 72-qubit machine making it the highest qubit machine in the world. I am confident as you are reading this that 72 has been surpassed. Google Research has a repository of information at https://research.google.com/pubs/QuantumAI.html - but be prepared for some highly technical information as well as more high level examples.

IonQ (ionq.co) is building the world's first fully-expressive, full-stack quantum computer based on trapped ion technology. IonQ's trapped ions represent one of multiple approaches being explored to power a quantum computer. IonQ believes trapped ion technology, which uses lasers to cool and isolate individual ions, will prevail because trapped ions are identical, more stable, can be better controlled, and are therefore likely to scale with better performance and greater predictability.

In February 2018 QuTech (qutech.nl), a Dutch quantum research and development company, created a programmable 2-qubit quantum processor on a silicon chip. They published their work in the magazine Nature - https://qutech.nl/programming-silicon-quantum-chip/

I was curious if the U.S. government sees the same extraordinary opportunities for quantum computing as the tech giants and I do. So, I did some research. This document buried deep within the Department of Energy web site shows it is very much on their radar.

National Strategic Computing Initiative
Computing Beyond Moore's Law Ceren Susut Advanced Scientific Computing Research (ASCR) Ceren.Susut-Bennett@science.doe.gov December 20th, 2016

National Strategic Computing Initiative
Strategic Objectives

1. Accelerating delivery of a capable exascale computing system that integrates **hardware and software capability to deliver approximately 100 times the performance of current 10 petaflop systems** across a range of applications representing government needs.
2. Increasing coherence between the technology base used for modeling and simulation and that was used for data analytic computing.
3. Establishing, over the next 15 years, a viable path forward for future HPC (high Performance Computing) systems even after the limits of current semiconductor technology are reached (the "post-Moore's Law era").
4. Increasing the capacity and capability of an enduring national HPC ecosystem by employing a holistic approach that addresses relevant factors such as networking technology, work-flow, downward scaling, foundational algorithms and software, accessibility, and workforce development.
5. Developing an enduring public-private collaboration to ensure that the benefits of the research and development advances are, to the greatest extent, shared between the United States Government and industrial and academic sectors.

ASCR plans a Quantum Testbed Facility that will provide the research community with access to early-stage quantum computing devices. The program's goal is to evaluate the utility of quantum computing to advance scientific questions of relevance to DOE to the extent that quantum computing proves value for DOE and facilitates the technology development required to provide production-quality quantum computing resources in the post-Exascale timeframe

The facility will:
- Advance quantum computing for science by allowing researchers greater access to early-stage technology
- Lead insight into how to assemble better systems for quantum computing and simulation
- Adopt the most advanced existing systems of qubits; adapt as technology evolves
- Complement quantum computing investments in NNSA and at other Federal agencies
- Address identified impediments to progress in quantum information science by:

Bridging institutional and disciplinary boundaries

Training a skilled quantum information workforce

Working synergistically with the emerging U.S. quantum computing industry.

In summary, it is important to keep quantum computing on your radar. Eventually, all computers will be based on quantum and what is impossible today will be 'easy peesey' for the next generation of computing technology. Start preparing for your business and career future now.

Step 3

Applying New Technologies to Create a Better and Smarter World

I want to get your idea juices flowing to the max! In this section, you will learn how the emerging technologies we discussed in the previous section are being applied to the way we work and play.

I will introduce you to many innovative companies leading the way and digitally transforming our world by applying using mindset, ideas and technology. I am confident you will be shaking your head in amazement at some of the solutions currently available and others on the horizon.

In some cases, you may want to leverage their solutions in your business or personal life. In other cases a light bulb may go on and you may decide you have a better solution and will build it.

I am not endorsing any companies, rather I am providing a starting point for you to learn more.

For your convenience, up-to-date links to all companies and products discussed in this section plus new golden digital nuggets are listed on DigitalSuccessBook.com/links.

The Smart Home
and Why It Matters

One of the hottest areas of digital transformation is creating the smart home. Reinventing the home and its impacts on quality-of- life are some of the most valuable aspects of emerging technology. Life changing products that were inconceivable just a few years ago, are now available and affordable to everyone.

We spend a lot of time at home. It is central to our lives. It is where we and our family share dreams and challenges. We want our homes to be as comfortable as possible. Smart devices will do that for us and a lot more.

As I write this there are millions of people facing heart wrenching decisions about putting a relative in a nursing home, assisted living facility or bringing in an aide. This can be life changing for the person impacted and for family that must make that decision with them and sometimes for them. It is a disruptive event in its most human form. Nursing homes are also very expensive. They can easily deplete the life savings of middle class families.

For the first time, there may be digital options that can provide 'live-at-home options. This excites me.

When my father-in-law faced the prospect of going into a nursing home in his early nineties I was against it. My personal feelings are that nursing homes should be used only if necessary. Fortunately, 'necessary' is rapidly becoming 'less and less' of an issue as new technologies become better and better solutions. Someday no one will go into a nursing home, but we are not there yet.

In 2010 when we had to make a decision about where my father-in-law would live I put together a workable solution for traversing stairs and where he would sleep but there were still many issues, such as making sure he took his medications as designed and monitoring his health. If we had to go out we were concerned about him not being able to turn lights on and off, set the control on the thermostat, raise and lower window shades, go to the kitchen for food and drink, seeing who was at the front door and many more everyday things we take for granted, but are significant challenges for others.

I am happy to report that between then and now technology breakthroughs have made living at home more practical and comfortable. As you read the rest of this chapter and the next one on 'Smart Opportunities to Help People with Disabilities', think about opportunities on multiple levels:
- How your organization can leverage these technologies or create new ones to help people live more fulfilling lives
- How you can repurpose similar smart devices in your office
- How you can repurpose similar smart devices to simplify your life and become more productive

We are still in the pioneering stage of building the smart home. You may think of new types of devices that consumers need or ways to retrofit your current devices to be smart. In the 'Playing, Prototyping and Profiting' chapter I referred to how I used a simple smart plug to control a standard lamp from an app and with verbal commands from Google Home. So, it is not always a requirement that the end device be smart if putting a smart device upstream enables you to meet your goal.

Below I describe a few of my favorite ways to make a home or business smart. It would take an entire book to describe all available smart devices for the home. As I mentioned at the top of the book any object can be made smart and provide new value. Anything is possible, so let your imagination run wild!

Smart Home Assistants: Chances are you already have one of these or you will have one in your near future. This category of IoT devices has taken off like a rocket. Leaders include Google Home, Amazon Alexa and Apple Siri.

They are important as they usher in a new verbal way to interact with technology. Someday typing emails, text messages and search commands will be a relic of the past. In the Artificial Intelligence chapter, we discussed the sometimes-comical responses from these devices. They will become more intelligent leveraging big data and AI in the cloud. In the coming years, they will approach perfection. These devices can be transformative for people who are not able to use a keyboard.

You can train them to learn your voice and your preferences. For example, if my wife asks, 'Hey Google - what is my mobile phone number?', Google Home will say her phone number. If I ask the same question, Google Home distinguishes my voice and provides my number. Of course, this is not a valuable use case but it can be when applying it to schedules, travel itineraries and more complex use cases.

Mobilizing smart home assistants will be the next step. The big digital players are working on this in their labs. I predict 2019 will be a watershed year for home robots that have built in digital assistance, understand verbal commands and the ability to do physical tasks for us because we do not want to do them or we cannot do them.

Another advancement will be to optionally bypass verbal questions and commands and perform direct brain-to-machine or brain-to-brain communications. You may think this is ridiculous but it is currently being explored in research labs and it can already be done wearing inexpensive head gear. I will provide examples of such products and services in the next chapter on helping the disabled.

Smart Thermostats: Nest (nest.com) is a born digital company that came from nowhere to disrupt Honeywell which had owned 30% of the U.S. market for thermostats. **Tony Fadell**, a founder of Nest is a brilliant technologist, inventor, designer, entrepreneur and investor. He previously worked closely with Steve jobs in was instrumental in growing Apple. He realized the thermostat niche was 'ripe' for the taking. Having heard Tony interviewed by Emily Chang and others, I consider him the LeBron James of technology!

For the past hundred years, a thermostat would be set to deliver a static temperature whether anyone was home or not. Energy was wasted and people paid needlessly high electric bills. The incumbent thermostat leading companies did not anticipate Nest's impact on their space. We discussed earlier in the book the importance of understanding disruption from not only within your niche but also from tangent or distant niches. Are there disruptors that have expertise and technology they can port to your space to leap beyond you and take your customers? Do you have technology and expertise you can use to own other niches? If not, you can probably pull in outside expertise to augment your in-house experts.

The experience Tony developed partnering with Jobs at Apple shows in the Nest product line. Each Nest product is sleek, simple to install and use. It combines sensors, connectivity and data analytics to create great customer value. The return on investment is clear and impressive. Saving money, energy and being a sustainable device creates customer happiness across the board.

A smart thermostat learns your preferences. Set it a few times and it uses data, AI and predictive analytics to learn your preferred temperature when you get home or when you have breakfast. You can even have it monitor your location so it knows when you are close to home. This can help you avoid needlessly using additional energy and spending money by turning up the heat or air conditioner at precisely the right time. You can control the device(s) from an app or by voice commands from your smart home assistant.

Nest wisely created a platform in which you can integrate non-Nest products. The difference between creating a product and a platform can be huge. We discuss platforms in more depth later in the book. For example, Phillips smart bulbs and Whirlpool washing machines can be seamlessly plugged into the Nest platform. Hundreds of other 3rd party products are listed on the Nest web site and more are added every week. Every product that is added increases the value of the platform. This is known as the network effect. Multiple smart devices interacting on a platform is a thing of beauty and can be life changing.

Although Tony Fadell was an integral part of Apple, Nest was acquired by Google, not Apple, in January 2014 for 3.2 billion dollars and is now a Google subsidiary.

Smart Video Cameras: I have been using a smart video camera for a few years and enjoy it. I use it to monitor my house when I am not home through an included app. Nest and many other vendors provide these. My smart device costs well under $100 and streams video to my mobile app. I can use the app to rotate the camera 360 degrees to get a view of my living room, dining room and foyer from many angles.

The camera also monitors for motion detection which is valuable to monitor secure areas that should not have movement. If motion is detected, I instantly receive an email and SMS message. I have also attached this to a commercial mass notification system for work to monitor secure areas. It is easy to do this using a data connector such as Zapier or IFTTT or with an API.

I also use the camera for fun to watch my dog when I am away. I thought it would provide a lot of laughs but I learned he sleeps much of the time or waits at the front door for my wife and I to get home. I can also see if he is being naughty.

Smart Lighting: Turning lights on and off using an app or by voice is easy. I have done both. I detailed the step-by-step process of setting this up in the chapter, 'How to Digitally Play, Prototype and Profit'. My family has not used a light-switch in my living room since we went digital. For people that have difficulty physically moving to control the switches this becomes life-changing.

Phillips and TP-Link are two leaders in the smart lighting space. They sell simple smart bulbs and higher priced ones that can control the hue and color of the light depending on a scenario.

Phillips is a great example of a company that stayed relevant instead of getting disrupted. They were a leader in lighting using incandescent bulbs for 100+ years. Through research and big data, they clearly understood the incandescent bulb business would be disrupted by cheaper / cleaner LED's and digital bulbs. They retooled their capabilities and business model and transitioned successfully to the future. They are now surviving and thriving.

Smart Window Shades: A simple voice command to your digital assistant can control opening and closing smart window shades. The actions can even be automated to trigger on a predetermined schedule. A fun and not too difficult project would be to integrate the smart shades opening and closing schedule with an online database of sunrise and sunset data. When sunrise is detected the shades automatically go up and at sunset they go down.

Once again, it is a digital win-win scenario that saves time, energy and is radically cool!

Smart Door Locks: Nest and Yale Lock have developed a tamper-proof, key-free deadbolt that connects to the Nest app. You can lock and unlock your door from anywhere. You can give people you trust a passcode, instead of a key. So, if a babysitter comes over you could let the person in from your app. In the future, this type of device may integrate with home deliveries from food and retail companies.

DoorBot (ring.com): As with many great inventions this one solved a problem for the inventor, **Jamie Smirnoff**. Inventors and software entrepreneurs seem to invent a lot of things in their garage. Jamie could not hear his front doorbell from his garage workspace so he hooked a wifi camera to an app which enabled him to see who was at the front door from not only the garage but anywhere in the world. The product provided a great deal of value to his family so Jamie formed a little company named DoorBot to see how far he could take his idea.

With a few employees and not much money his big break came when he was invited on Shark Tank in 2013. I remember the episode vividly. Most of the Sharks were not too impressed with the product and some felt Jamie overvalued the company. He did not get an investment from them.

After the publicity on Shark Tank, DoorBot steadily grew and the name was changed to Ring, Amazon took an interest in the product and invested in Ring. Perhaps, this was a strategic move for Amazon to Google buying Nest.

Things went well and when Amazon decided to move more aggressively into the smart home market they acquired Ring... for 1 billion dollars! Jamie is a very rich man. He owes some gratitude to Shark Tank for the publicity and the Sharks for not investing.

Roomba - there are millions of these iRobot (irobot.com) smart vacuum cleaners in homes world-wide. They combine a powerful cleaning system with intelligent sensors. They can seamlessly move through a home, adapting to the surroundings.

Imagine you will be entertaining friends and you are in a rush to pick up lunch, just tell Roomba to clean the floors and go on your carefree way.

You can create cleaning preferences and even get detailed cleaning metrics reports on your mobile phone including completed cleaning cycles, coverage area and duration. As with many other smart devices you can control Roomba with voice commands through your smart digital assistant.

Smart Water Detector: Unfortunately, I know from personal experience that water damage is expensive to fix and sometimes the full impact never goes away. The goal of the **D-Link Water Sensor** (http://us.dlink.com/products/connected-home/wi-fi-water-sensor/) is to let you know when water is detected before it's too late. You receive an alert on your mobile device when a water leak is detected.

Smart Kitchen Appliances - there are many and the list is growing. The ones you might be familiar with are smart coffee makers and refrigerators. Allow me to describe a few less common ones that provide value:

WeMo has developed the **Crock Pot Slow Cooker** (http://www.crock-pot.com) that enables you to adjust cooking settings from your mobile device. It lists for under $100.

The Perfect Company (https://makeitperfectly.com) is a leading developer and innovator of connected kitchen products. They have developed some interesting smart devices including Perfect Bake, for baked goods; Perfect Drink, for the perfect drink; and Perfect blend, a smart blender. Their smart scales and guided recipe apps enable easy, seamless, beginning-to-end food and drink preparation at home.

They strive to make the kitchen experience free of measuring utensils, and produce delicious food and drinks. In addition, their app keeps track of users' nutritional intake for every recipe.

The devices in this chapter can make your life easier. Many can be life-changing for elderly and disabled people. I cannot wait to share technologies and specific devices that can help people in so many ways in the next chapter. So, let's do it!

Smart Opportunities to Help People with Disabilities!

Worldwide, approximately one billion people have a disability, as stated by the World Health Organization. Helping the disabled is a broad category. Digital technology will change the lives of millions of people with disabilities. Soon most disabilities will only exist in our memories.

In my research for this chapter I came upon many stories of people that suffered catastrophic events such as strokes and accidents and many were effectively, 'locked in their body'. They did not have the ability to move. In some cases, they only had the ability to move their eyes. Many other people lost their sight or ability to walk. Digital transformation can transform their lives, which is why this chapter touched me the most of any in the book.

Whether you have a disability or not this chapter is about thinking without limits. For example, it has been estimated that the BCI (brain-computer-interface market) can grow to 14 billion dollars by 2024. Although the focus of this chapter is centered on helping the disabled, these technologies can be applied to many other use cases and verticals. They may help you in your personal life, business or in the assistance of someone you know. Think big and out-of-the-box as the innovators in this chapter have. Anything and everything is possible.

My hope is that everyone reading this chapter understands how life-changing technology, innovation and creativity can be. If there is a need you have and you cannot find a place to get help with it contact me and I will try to assist you.

Perhaps, your business can aid people in some way. Happily, large corporations such as Microsoft, Dell and Citibank are using digital assisted technologies to help disabled people become assimilated into the workplace.

Many apps are available to assist people with disabilities. For example, **SeeingAI.com,** is a great free AI app from Microsoft that can be life-changing for people with disabilities. It is designed for the low vision community and uses your mobile device camera to describe people, text and objects.

In the spirit of experimentation, it was built as a research project at a hackathon in July 2017 and has millions of downloads. I saw a demo and immediately downloaded it. If you know anyone you think can benefit from it, please spread the word. I would also be interested in feedback if it was helpful.

Be My Eyes (bemyeyes.com), a winner of the Google Play Store 2018 Best App Awards, is a free Android and IOS app that connects blind and low vision people with sighted volunteers and company representatives for visual assistance through a live video call. The sighted person can assist the blind person with questions and tasks. The scope of questions is wide. For example, in a supermarket 'which cookies am I buying' or 'which color apples am I buying?' For public transportation, 'how many minutes does the screen say until the next bus arrives?' These are situations that many sighted people take for granted and the volunteers can be so helpful and comforting to people without sight.

There are many heartwarming stories of how volunteers have made life better and more independent for people using the platform. I am now proudly a volunteer!

AssistiveWare (http://www.assistiveware.com) has a mission to empower people through innovative technology. They have developed several assistive technology software products for Apple's Mac OS X, iPhone, iPad and iPod Touch.

In addition, at the request of their users, AssistWare collaborated on developing text-to-speech in children's voices. Nearly half of the users are under the age of 12 and unfortunately there were no genuine children's voices yet available. Users would use adult voices or artificially modified voices with raised tones that sounded like they had inhaled helium. This meant that most young AssistWare users had to speak in a voice they could not identify with and which seemed unnatural or implausible to their communication partners. The new text to speech voices helped children fit in better with their peers. You can listen to various children's voices from different countries here - http://www.assistiveware.com/innovation/creating-genuine-childrens-voices

A prosthetic hand can cost thousands of dollars. Children can quickly outgrow them as they get older. **e-NABLE network**, (enablingthefuture.org) which is a volunteer network, now makes it possible for many children to have a new prosthetic hand for a fraction of the cost by printing them with 3D printers. I suggest you visit their website and watch a couple of videos. I think you will shed some tears of joy. This is what enabling technology is all about!

Here is an idea, if you have 3D printers sitting idle possibly they can be used to print prosthetics to transform lives and make you and your organization feel good.

In addition to prosthetics, other types of medical devices produced by 3D printing include orthopedic and cranial implants, surgical instruments and dental restorations such as crowns. Life-saving internal organs such as windpipes have been printed by hospitals.

In 1986, a group of engineers gathered in a basement in Virginia and founded LC Technologies. Since its founding, the company has pioneered the development of eye tracking technology for over 26 years. Their Award-winning Eyegaze Edge® (eyegaze.com) has a long-standing reputation for accuracy, reliability, and ease-of-use. LC Technologies now operates in 44 countries.

400,000 people worldwide are effected by motor neuron diseases and multiple sclerosis affects 2.3 million people. The late Professor Stephen Hawking was effected by motor neuron disease but was able to lead a fulfilling life using technology that enabled him to use eye movement to control a computer.

Their eye-tracking systems, are hands-off, unobtrusive, remote human-computer interfaces. All of their eye tracking systems have consistent gaze point prediction and accuracy even when a subject falls outside the "norm". There are a wide range of applications but the first systems were designed to enable severely disabled individuals to communicate using only their eyes. To date, their technologies are now being implemented in research, national defense, gaming, virtual worlds, hospitals and many more areas. Possibly they could add value to your industry and customers.

How does the Eyegaze Edge speech generating device improve communication in the home? Families that contact LC Technologies Inc. for eye gaze communication solutions come to them, as they are experiencing communication declines in the home with their loved ones. Since communication is a shared activity, not only do these declines impact the individual who has lost his voice, but greatly impacts relationships with family members, caregivers and friends as well.

With declines in speaking ability, before receiving an eye gaze communication device, an individual often experiences:
- Failed attempts to communicate their intended messages
- Abandonment of ideas
- Reduced participation in daily routines
- Decreased sense of connectedness with the world around them
- Increased frustration, despair or sadness
- Dependence upon communication partners to interpret communication attempts
- Dependence upon communication partners to anticipate daily wants and needs

Eyegaze Edge allows the individual to once again:
- Type and create messages using their own voice
- Advocate for themselves
- Participate in family decision making
- Agree or disagree with someone
- Express their wants, needs and desires
- Communicate in emergency situations
- Direct the behavior of caregivers
- Engage people in their community
- Work from home
- Use a telephone, the internet and send text messages
- Participate in leisure activities
- Participate in support groups
- Nurture family relationships through communication

eSight (https://www.esighteyewear.com/) has created digital technology to help the visually impaired see the world, and change it too. eSight restores functional sight, and allows the visually impaired to see faces, read, work, study, and participate in virtually any activity.

Throughout the book, I tried to communicate that if you dream it, you can make it happen. eSight says it perfectly on why they created their life-changing product: 'Our movement began when our founder dared to dream of a better world for his two legally blind sisters. He wanted nothing more than to see them be truly mobile and independent — free of the limitations that came with blindness, and the many single-purpose assistive devices that had become the norm.'

I was further moved by their 'What we deeply value' mission statement:

'Our core belief is that Everyone Deserves to See. We believe in universal access to sight and the experiences, both essential and beautiful, that it enables. This has inspired our iconic global commitment to Make Blindness History by 2020, now that a technology as transformational as eSight exists.

We will turn the tide on the sky-high unemployment rates, educational challenges, and discrimination that have followed the legally blind for far too long.'

On their site, they have a short, 'Want to know if eSight will change your life?' express screening to find out if eSight can be of help to you.

Handisco (handisco.com) is the inspirational story of a French start-up that partnered with the digital giant, Cisco, to build Sherpa, an innovative solution that can benefit millions of people. This type of combination can have powerful results as demonstrated below. I also suggest this video that shows the product in action and presents a high-level road-map of how a startup goes from concept to market (https://newsroom.cisco.com/video-content?type=webcontent&articleId=1863824)

The following is used with the permission of http://thenetwork.cisco.com/

With support from Cisco, French start-up Handisco developed an IoT walking stick to help visually impaired people navigate around their communities.

Imagine trying to navigate a busy town without the benefit of sight, using only a grey cane to help you safely get around. A couple of college students saw the problem and wanted to come up with a solution to make getting around town a little easier for the visually impaired.

They came up with a design for an IoT enabled device that straps onto a cane and is Bluetooth enabled.

Soon after, they created their company, Handisco. "Handisco comes from the French prefix Handi, which is the French word for disabled people," says Florian Esteves, one of the company's co-founders.

Through their university, Handisco's co-founders heard about a competition sponsored by Cisco, called the Switch-Up Challenge. They entered two years ago, and won not only seed money to get their company off the ground, but also mentoring from Cisco employees. "It was good help for us because we were rookies," Esteves said. "We are engineers so we do not know so much about strategy."

The web site is in French but you can use your browser translate feature or other languages. As of August 2018, Sherpa works everywhere in metropolitan France and in French speaking Europe (Belgium, Luxembourg, etc.). Assistance with public transport is available in the 40 metropolises covered. New cities are coming and may already be available when you read this.

The young entrepreneurs called their creation Sherpa, after the Himalayan mountain guides. "For us, it's kind of the same thing in the city," Esteves said. "The goal of this product is to help people to achieve more and more things in the city."

The Sherpa works by using GPS and IoT technology, and can communicate and connect with a pedestrian light to let someone know when it's safe to cross the street.

The Sherpa can also guide someone with step by step directions and can give them detailed information on bus routes and pickup times.

"We have been supporting Handisco for a few years now through their incubation phases and now indeed they're ready to go to market," said Alain Fiocco, a Cisco Senior Director in Engineering.

Bristol Braille Technology (http://www.bristolbraille.co.uk) is a Social Enterprise working from the Bristol Hackspace. They are building a revolutionary and affordable Braille e-reader for blind people called the Canute 360, designed with and by the blind community. It has been described as a sort of braille Kindle. The product was in the testing phase in August 2018 with a goal of launching later in the year.

Better Hearing aids - For many decades hearing aids were analog. When they receive sounds they amplify them. Foreground voices and background noise both get amplified.

In 1995 hearing aids were disrupted by going digital. A digital hearing aid includes a tiny microprocessor that converts sound waves to digital signals. This enables them to enhance the sound and filter out unwanted noise. There will be many more benefits from digital hearing aids as our world continues the digital explosion.

Autism - One out of every 68 children in the United States is on the autism spectrum per the CDC. Digital tools have been reported to help children with autism. I have seen emotional reports from parents on how technology changed the lives of their children. **Autism Speaks** has an informative page focused on technology and feedback from visitors at https://www.autismspeaks.org/family-services/community-connections/technology-and-autism.

Bypassing Damaged Body Parts with Digital Substitutes
Second Sight - Argus II Retinal Prosthesis System (http://www.secondsight.com) provides electrical stimulation of the retina to induce visual perception in blind individuals. It is indicated for use in patients with severe to profound retinitis pigmentosa.

Here is how it works - a miniature video camera housed in the patient's glasses captures a scene. The video is transmitted to a small patient-worn computer where the video is processed and transformed into instructions that are sent back to the glasses via a cable.

These instructions are transmitted wirelessly to an antenna in the retinal implant. The signals are then sent to the electrode array, which emits small pulses of electricity. These pulses bypass the damaged photo-receptors and stimulate the retina's remaining cells, which transmit the visual information along the optic nerve to the brain, creating the perception of patterns of light. Patients learn to interpret these visual patterns with their retinal implant.

The company that developed the Argus II System, Second Sight, is testing a new device called Orion which does not require glasses. In January 2018, they announced that the first human patient to receive the Orion cortical visual prosthesis system was implanted with the device, as part of a feasibility clinical study.

Bypassing the damaged retina altogether
Second Sight called the first-in-human implant of the Orion a significant milestone, and a critical step forward in its development of devices that it believes could potentially treat nearly all forms of blindness.

The Orion implant applies a similar principle but bypasses the eye altogether. It is implanted into the surface of the brain itself near the visual cortex, in which it stimulates directly with the signals received from the wearable video camera after processing.

This architecture avoids the use of the damaged retina and the optic nerve altogether, in an analogous fashion to the principle behind cochlear implants used as a treatment for deafness, and could therefore be of value in the treatment of multiple forms of vision loss. It does, however, involve more invasive surgery than an epiretinal implant. In addition, the precise workings of the brain's visual cortex are complex and still not fully understood.

Mind to Internet Interface: Soon you will be able to communicate with the Internet and physical devices directly from your brain. There will be no keyboard, touchscreen or voice commands are required! Your thoughts will interface directly with systems. You can request a ride-share, change the channels on your TV or Google anything you can think of.

For people with certain disabilities this can be life-changing. Imagine being bed-ridden and unable to move. Even if you can move your eyes, going straight to the brain is much faster, more natural and less tedious than eye movement to input letters or symbols or to control a stylus with your mouth or foot. There is also experimentation on combining Brain Computer Interfaces with virtual reality and artificial intelligence to improve medical research by creating accurate simulations of the brain. But, is all of this pie-in-the-sky? You be the judge:

The goal of an interface to the brain is not new. Researcher William Dobelle developed a prototype that was implanted in

1978 into a man blinded in adulthood. A 68-electrode single-array BCI was implanted onto the visual cortex and succeeded to give the patient the sensation of seeing light.

The system naturally was cumbersome, as the technology of the period was nothing like the present. It included cameras mounted on glasses (sort of like Google Glass) that sent signals to the implant. Amazingly the implant allowed the subject to see shades of grey, although in a limited field of vision and the subject had to be hooked up to a mainframe computer. All this hardware has now been miniaturized to fit on a chip.

MIT Media Lab is an amazing place. They are a modern-day Xerox Park, responsible for planting the seeds of the future. Many of the magical technology advances we now take for granted, such as autonomous car routing and the first eBook with full font rendering in the 1970's, originated at MIT.

In April 2018, an episode of CBS's 60 Minutes focused on the MIT Media Lab. I learned of a jaw-dropping thought transfer project at MIT. Scott Pelley, the interviewer, and I were both amazed and I knew instantly I had to mention it in my book.

Arnav Kapur is the project leader of **AlterEgo**. It consists of a wearable device that intercepts electrical signals, which the brain normally transmits to vocal cords. Instead, it sends the information to a computer. The device has a safety feature so it does not reveal a user's private thoughts. It must deliberately be activated by "internally vocalizing," (talking to yourself).

When the thoughts are sent to a computer the words can be displayed on a screen, vocalized or sent to a search engine as a question or to a device as a request to perform a task.

The computer then sends the response back as vibrations transmitted through the user's skull and are transferred to the inner ear which the user hears internally.

Amazingly, Arnav and his team had only been working on the project for one year as of the airing TV episode. The implications for people with disabilities is profound.

This is a link to beyond the scenes of the AlterEgo part of the episode: https://www.cbsnews.com/news/mit-media-lab-where-tomorrows-technology-is-born/.

Below is a link to the entire video of the episode that includes AlterEgo at the beginning and additional mind-boggling projects, including a person controlling a prosthetic with his thoughts as part of the incredible work of Professor Hugh Herr, who leads an advanced prosthetics lab. I highly recommend this inspirational video: https://www.cbsnews.com/news/mit-media-lab-making-ideas-into-reality-future-factory/

Regina Dugan directs Facebook's secretive research and hardware lab **Building 8**. Previously she ran the U.S. military's research lab, DARPA, and then worked at Google. Facebook is building what it calls a "brain-computer speech-to-text interface" technology that is supposed to translate your thoughts directly from your brain to a computer screen without any need for speech or fingertips.

The idea is that this technology will be able to take what you are thinking to yourself in silence, using non-invasive sensors that can read exactly what you intend to say, and turn it into readable text.

Two proven entrepreneurs are taking a crack at making BCI a reality using brain implants. Both are in the beginning stages as of 2018 but because each of these people are 'finishers' (they do what they say) I wanted to at least mention their newborn companies. Whereas, skull-caps commonly used for BCI are termed 'non-invasive', this technology is termed 'invasive', as it must be implanted in the brain. People are already having other types of chips implanted as I described earlier, so someday brain implants might be as standard as other implants. That said, it must be regulated very carefully.

If he is not changing the world enough, Elon Musk, CEO of Tesla and SpaceX, is launching a new company focused on brain-computer interface technology, called Neuralink (neuralink.com).

Bryan Johnson launched Kernel (kernel.co). He previously founded Braintree which was acquired by PayPal for 800 million dollars. His idea is centered on tiny brain implants to expand the bounds of human intelligence.

As futuristic as all the above sounds, I am confident we will surpass all of it by 2040. As you have learned in this chapter we have already figured out how to encode brain analog signals to digital so anything is possible. We can transfer thoughts and control IoT objects. We can transmit questions to Google and receive answers.

So why couldn't we take it a step further and do brain to brain thought transfers? Possibly, in the beginning it would not be direct B2B (brain to brain) but through a secure (possibly tied to a DNA marker) interface that accepts the thought, matches it with a B2B directory and sends it to a receiving party that 'friended' the sender. The receiver would consume the thought using a non-invasive or invasive chip.

Along the way, I predict a software brain operating system (BOS) and carefully regulated standards will be developed to make it easier to build safe and secure devices and applications. Medical professionals will work side-by-side with software engineers to free people from the constraints of disability.

People that do not have disabilities will also benefit from brain to brain thought transfer. There will be no more hardware to carry around, recharge and possibly lose. Just a miniature cell-size interface that can be powered by the body and last for decades or a lifetime. Imagine, the knowledge of the Internet only a thought away. Say goodbye to charging your clumsy devices every day or typing commands! Our ancestors will laugh that we carried eco-unfriendly mechanical devices everywhere we went.

I am optimistic and elated that brilliant innovative people and digital technologies are bringing us closer to eliminating disabilities and enabling far reaching communication technology.

Smart Personal Transportation - Faster and Better

Moving people from point A to point B is changing dramatically for the better in the digital age. Autonomous cars, trucks, flying taxis and the Hyperloop are all in our personal transportation future.

Electric and hydrogen powered vehicles will reduce fossil fuel pollutants. Our air will be cleaner which will improve our health. The global infrastructure of charging stations is growing and electric batteries are becoming more efficient. These batteries can serve double duty in a power outage as tiny power stations to keep the lights on in our homes.

Self-driving automobiles are being tested on public roads and will soon significantly reduce the thousands of unnecessary accidents and fatalities caused by human error and by drunk and distracted driving. In 2017, 40,000 people died in car accidents in the U.S. and millions were injured or disabled. Worldwide 1.3 million people died in car accidents. That must stop!

Google's **Waymo** is the leader in the autonomous vehicle space. As of mid 2018, they had completed over 6 million self-driven miles on public roads. There is a lot of competition from practically every car maker on the planet and services such as Lyft and Uber. Competition and ingenuity will 'drive' our commuting experience to a new level sooner than many people think.

Waymo has purchased and is testing thousands of autonomous Chrysler Pacifica minivans as part of their EarlyRider program. They have also partnered with Jaguar to add 20,000+ I-Pace SUV's to their self-driving taxi testing program. In May 2018, they announced the addition of up to 62,000 more Fiat Chrysler vehicles to their program. I see this as a fruitful partnership between a digital born company and an industrial age company boldly moving into the digital age.

Smart car advances are already improving the safety and comfort of driving. My wife's 2017 Mazda is more computer than a traditional car. The side mirrors sense cars to the left or right and in the blind spot to avoid lane changing accidents. The back-up camera can be life saving.

The engine is electronically maintained for better performance and the tires communicate pressure related issues so we can adjust and avoid a blow-out or unnecessary drag on the car do to under-inflation.

After thorough testing is completed, the first wave of self-driving cars on our roads will have a feature for human drivers to take the wheel but eventually that will go away. Waymo and others are experimenting with completely autonomous cars with no steering wheel.

Early metrics demonstrate the fear of not having control goes away quickly and we are safer and more secure with a computer at the wheel. In addition, our cars will use their sensors and AI driven software to get smarter every mile they drive. Eventually accidents will be a thing of the past and a fender bender minor accident will be an anomaly.

Workforce productivity and enjoyment will increase significantly when people can sit back and not worry about driving or sitting in traffic. They can work, eat lunch, read a book or watch a movie. The term distracted-driver, which is often used in conjunction with tragic accidents, will no longer exist. Perhaps, cars and trucks without humans at the wheel can also eliminate vehicles being used as weapons.

Traffic will be reduced as autonomous cars will eventually communicate with each other and will be able to drive closer together without crashing. For that to happen we will need the implementation of autonomous car hardware, software and communication standards.

Artificial intelligence will discover patterns in traffic and communicate with vehicles either directly to control the engine or through future upgrades to Waze type apps. Today Waze has networked drivers to provide useful information such as debris on the road or traffic jams. The next step is for each car to be a node in the network.

Ride sharing has already changed our lives. You are familiar with companies such as Uber and Lyft but there are many other companies prospering worldwide offering ride-sharing and add-ons. For example:

GrabTaxi (https://www.grab.com/sg/) rules the streets in Malaysia, Indonesia, Thailand, Vietnam, Philippines, Myanmar and Cambodia. Grab has morphed beyond motorcycle ride-sharing into many related niches such as package delivery and GrabPay digital payments. They are creating value every second of the day in multiple ways.

I watched Grab's CEO Anthony Tan interviewed on Bloomberg TV's High Flyer show. Each episode features a different tech leader interviewed on a ferris wheel. You can watch episodes online (https://www.bloomberg.com/video/high-flyers/3/). Anthony is a brilliant visionary who can execute a big plan. I would not bet against him.

EasyTaxi (http://www.easytaxi.com/) rules South America. As of 2018 they have 20,000,000 riders in 170 cities and 12 countries.

DiDi Chuxing (http://www.didichuxing.com/en/) is chugging along on all cylinders in China. They have a ride-sharing app the enables 400 million riders in 400 cities throughout China to enjoy private cars, taxis and designated drivers. But wait - there's more! Didi Chuxing also offers DiDi Chauffeur, DiDi Bus, DiDi Minibus and DiDi Car Rental. Their site has a wealth of information. I love the way they describe who they are:
"At its core, DiDi is a data company that relies on AI capabilities to advance new mobility solutions and breakthroughs in transportation technology."

The conventional taxi industry has been disrupted and it is sad how out of touch they were to the disruptors that now own their space. Their feeble counter-punch frustrates me, as a lot of cab drivers and medallion owners are now suffering.

Unless the major automobile companies see the haymaker around the corner, they are going to be just as seriously disrupted as the taxi industry. Automobile ownership is going to be decimated for many reasons;

Green is smart and automobiles are wasteful. Not only because of the fossil fuels they burn that are killing us through pollution and yes, that will be reduced when they are powered by electricity or hydrogen, but more importantly because they are expensive and on average sit idle 94% of the time. Plus, a large portion of the other 6% of the time they are used is for trips of 4 miles or less!

Millennials are not into cars nearly as much as baby boomers and prior generations. Cars are not a status symbol for them or me. They would just as soon click on an app and a ride-sharing service car appears in minutes than dole out hundreds or thousands of dollars a month to buy and maintain a car that is a big paper-weight most of the time.

What will also change is the current need for ride-sharing drivers. Ride-sharing companies will eventually replace all their drivers with self-driving cars.

Eventually owning and driving a car, for most people, will be a relic of the past. People that do own an autonomous car will use them for their personal enjoyment and when they are at work or on vacation they will add their vehicle to a ride-sharing network so it can pick-up riders and generate revenue while the owner is having fun.

Fast electric car charging:

Is it fantasy to imagine being able to fully charge an electric car in the time it takes to boil a pot of water? Well, soon a super-fast 5-minute electric car charge may be a reality. That would transform the value of owning an electric car. For me, it would weigh heavily in my decision to buy one.

Often, these types of cascading effects are overlooked, which is why they can be so valuable to companies and investors that figure them out in advance.

Charging electric car batteries has historically been the weak link. Although they are more sustainable than using fossil fuel gas, the difference in the extraordinarily long time it took to charge compared to a quick gas purchased caused buyer friction. This problem was further magnified, as batteries only partially charged were impractical for long drives.

StoreDot (store-dot.com), an innovative Israeli company may change the equation. They are a nanotechnology materials pioneer known for their super-fast battery charging technology. In 2017 they demonstrated their five-minute super-fast car battery charger at the CUBE Tech Fair in Berlin.

Their Flash Battery technology can make charging any electric car as quick and efficient as filling a gas tank, as it only takes five minutes to reach a full charge that can keep certain electric cars going for 300 miles. In May 2018 BP invested $20 million in StoreDot.

"Fast Charging is the critical missing link needed to make electric vehicles ubiquitous," says Dr. Doron Myersdorf, Co-Founder and CEO of StoreDot. "The current available battery technology dictates long charging times which makes the EV form of transport insufficient for the public at large. We're exploring options with a few strategic partners in the auto space to help us boost the production process in Asia and reach mass production as soon as possible. "

SmartDot's proprietary technology, inspired by nature, can be optimized for multiple industries. In May2018 TDK Japan invested in StoreDot and signed a joint development contract to try to apply StoreDot technology to smart phones, earphones and smart watches.

In addition to all the advances in automobile transportation virtual and augmented reality will have an impact by reducing the need for humans to physically move from point A to point B. Virtual meetings and experiences will be so real in the near future much of the resource heavy business travel we do out of necessity will be reduced. Your hologram will instantly be teleported across the globe. Your hologram can attend meetings in 5 continents in one day without you leaving your home office!

Flying Cars and Taxis Fantasy Becomes Reality
The flying car, which many of us dreamed about while watching The Jetson's in cartoons, is nearing reality.

What NASA is saying and doing: "A long-standing futurist dream is that of an individual flying car: garaged at home, but able to make daily quick and easy jaunts of several hundred miles for work or play. While several companies may field airborne autos in the next several years, urban commuters could have another option: air taxis. Langley researchers have begun work on ways to make such
personal air mobility a reality. A key component will be

the inclusion of some form of autonomy.

Broadly defined, autonomy is the ability of an unpiloted vehicle to make its own decisions literally on the fly. To explore the rapidly advancing field, Langley's Autonomy Incubator assembled a team of civil servants, contractors and student interns with diverse expertise in everything from machine learning and vision, to human psychology, to mechanical engineering, to computer science and electrical engineering.

Other Langley autonomy work continues across an array of aerospace disciplines, including onboard machine learning for self-management and decision making; ways to sense and avoid obstacles; methods to maintain uninterrupted communication with other craft and with ground monitors; and means of instant response to internal and external hazards. Research is also being conducted into autonomous self-assembly of both in-space and on-planet structures in anticipation of humanity's push beyond Earth into the solar system."

In the private realm, Uber has already been exploring flying taxis. In February 2018 Uber CEO Dara Khosrowshahi predicted at an investors conference that Uber flying taxis will be commercialized and used widely within the next decade. It may seem like a boastful claim but during 2017 and 2018 Uber hired senior NASA and aerospace employees. In July 2018 indications seemed to be that Uber was moving up the goal of flying taxis to 2 to 5 years hence.

Kitty Hawk (https://kittyhawk.aero) is a company developing two types of flying vehicles - Cora and Flyer. As of this writing in August 2018, Google's Larry Page is an investor in the company.

Cora brings the airport to you. Their web site says, "Cora began as a dream. An air taxi so personal and so simple it could take the trips you make every day, the ones that define our lives, and bring them to the sky. Cora isn't just about flying. Cora is about the time you could save soaring over traffic. The people you could visit. The moments that move you."

Cora has a timeline of how they got to now: (https://cora.aero/milestones/). In August 2018, there were also several careers listed on their site.

Flyer seems more recreational for over water flying and perhaps it is a lot cheaper. Here is a video that shows it piloted by an aero-engineer. Their goal is that regular people will easily be able to learn to fly it: (https://youtu.be/mMWh4W1C2PM)

Aerbus is making great progress with their Vahana self-piloting vehicle (https://vahana.aero). They had a successful first test flight January 2018 and report behind the scenes progress on the Vahana web site. They have resources and skills and will be a leader in this space.

Aerbus Innovations (http://www.airbus.com/innovation.html) is very interesting. There are many areas of innovation and if you scroll to the bottom of the page there is an Open Innovation email form if you want to submit your own ideas.

Volocopter (https://www.volocopter.com/en/) is already in test mode with the goal of building the first manned, fully electric and safe VTOL (Vertical Takeoff and Landing) vehicle in the world. Their vision integrates air taxis into existing transportation systems and provides additional mobility for up to 10,000 passengers per day with a single point to point connection.

They have an outstanding video on their site of what it would look like as their vehicle flies through a city and lands on an extension platform on a skyscraper. It is the kind of feature functionality that would have been a highlight of a futuristic movie only a few years ago, and is now quickly becoming a reality. Plus, they have jobs and lots of additional interesting information on their site.

AeroMobil (https://www.aeromobil.com) has two concept flying cars - AeroMobile 4.0 STOL (Short Takeoff and Landing) and 5.0 VTOL. Their web site says, "AeroMobil aims to lead the industry with the most comprehensive door-to-door flying car solution covering the first-to-last mile in all weather conditions, for short, medium and longer distances up to 700km in a single journey."

Lilium (https://lilium.com/) is developing an electric vertical take-off and landing jet. You can watch the maiden flight on their web site. The idea is for it to

serve as an air taxi. You can summon it to a nearby landing pad. The capabilities of the sleek looking air taxi as described by Lilium will be impressive:
- 300 km range - London to Paris in one hour
- 300 km per hour speed - no emissions or air pollution
- Low noise - less than a motorbike

Lilium's career page, as of August 2018, had numerous jobs listed.

Pal-V (https://www.pal-v.com/) was taking pre-orders in 2018 for their flying car. They are expensive and not shipping as of August, but they say they will ship in 2019

It is too early to know which of the flying cars mentioned above will come to fruition. What is clear is that there will be flying cars in our future.

The Hyperloop - will move people and cargo between cities at very high speeds. Hyperloop reminds me of how mail was sent through a pneumatic tube in Manhattan, NY 100 years ago.

You sit in a pod and can leave anytime you desire. Unlike a train, there is no schedule you must worry about. It can run above or underground. Electric propulsion is used to gradually accelerate the vehicle to airline speeds of approximately 670 mph / 1080 kph, as it glides on a cushion of air using magnetic levitation. It is aero-dynamically efficient and 10-15 times faster than traditional rail travel.

As of September 2018, in my opinion there were three leading companies:
- Dirk Ahlborn and Bibop Gresta's Hyperloop Transportation Technologies
- Richard Branson's Virgin Hyperloop One
- Elon Musk's SpaceX Hyperloop / Boring Company.com
- All three companies are in the development and testing phases and are making significant progress.

Hyperloop One describes numerous hyperloop studies including Chicago, Dubai and Missouri on their site. They also have a terrific video in which you can visualize that you are riding in a hyperloop. Their goal is to have people and cargo hyperloops in service by 2021.

SpaceX Hyperloop (http://www.spacex.com/hyperloop) / Boring Company (Boringcompany.com) The Hyperloop system built by SpaceX at its headquarters in Hawthorne, California, is approximately one mile in length with a six-foot outer diameter. The Boring Company is focused on creating tunnels with hyperloop for long distances and loop for shorter destinations.

At press time for this book they had announced agreements for a hyperloop to be built in Chicago.

SpaceX Hyperloop has run a Hyperloop Innovators Competition for universities from 2015 through 2018. There is a video on the website of the 2017 competition. The 2018 competition took place on July 22, 2018 and focused on a single criterion—maximum speed. Additionally, all Pods had to be self-propelled. Hopefully there will be future competitions.

Hyperloop Transportation Technologies (HyperloopTT) (http://www.hyperloop.global). Although they are not as well-known as their competitors, I would not bet against this company. Researching their vision, aggressive plans and successes, it seems they can be a serious player. Using unique, patented technology and an advanced business model of lean collaboration, open innovation and integrated partnership, HyperloopTT is creating and licensing technologies.

They were founded in 2013. They have a global team comprised of more than 800 engineers, creatives and technologists in 52 multidisciplinary teams, with 40 corporate and university partners. They are headquartered in Los Angeles, CA with offices in Abu Dhabi and Dubai, UAE; Bratislava, Slovakia; Toulouse, France; and Barcelona, Spain. They have signed agreements in Ohio, Slovakia, Abu Dhabi, the Czech Republic, France, Indonesia, Korea and Brazil.

In 2018 they began construction of the first full-scale passenger and freight prototype system in Toulouse, France, and expect to deliver the first passenger capsule in late 2018. The capsule will be assembled and optimized in Toulouse, France, prior to use in the United Emirates.

Dubai is a leader in using smart technology and has been since 2008 with the creation of a fascinating experimental smart city named Masdar. Many digital transformation initiatives are under way in Dubai. On April 18th HyperloopTT and Aldar Properties signed an historic agreement for the world's first commercial Hyperloop system of 10km in the critical development area between Abu Dhabi and Dubai

Dubai is also testing flying taxis from the previously mentioned Volocoptor. They want to be the first city to have them in production.

The Khaleej Times reported in May 2018 that a trial 'bus on demand' service was launched in Al Warqaa and Al Barsha. Ahmed Bahrozyan, CEO of the RTA's public transport agency, said: "The 18-seater buses will operate on flexible routes and schedules, and bus drivers can know the service demand through the app to reach to the nearest point to their destinations. Through the MVMANT app, customers living in Al Warqaa and Al Barsha will be able to identify their locations and track the bus arrival times to their locations." Seems like a throwback to my son's 'Bus is Near' project.

Dubai is also implementing a network of driver-less pods. These systems will have 25 driver-less vehicles that can carry 24 passengers each. The first phase will run on special 'tracks'. Later the hope is to for them to operate with other traffic. The network is part of Dubai's Smart Autonomous Mobility Strategy and the hope is it will be operating by 2019.

Innovative minds will continue to develop new and clever ways to move people. Here are two more creative ways to move people.

Segway miniPRO (http://www.segwayminipro.com/) which I detailed in the Robots and Drones chapter is a form of personal transportation.

Organic Transit - Elf Pedal - Solar hybrid vehicles (https://organictransit.com/) - I have been following the Elf for years. They empower personal transportation and the extra benefit of exercise and sustainability. On their web site they say, "The ELF is a solar and pedal hybrid vehicle powered by you and the sun. "The most efficient vehicle on the planet", it is a revolution in transportation and gets the equivalent of 1800 MPG."

The ELF can achieve 20 MPH with electric assist, and 30 MPH with pedaling. The electric motor can transport you up to 48 miles with no pedaling! There are no carbon emissions, and complete independence from fossil fuels. Payload is between 350-550 pounds. The vehicles even have protection from the elements. They have a cool action video on their web site.

New technology is making anything possible. Let your imagination run wild and you can become the next 'people movement' titan.

Smart Supply Chain
Our Chain of Life

Businesses have many opportunities to automate and digitally enhance their supply chain processes using emerging technologies. Already we have seen costs reduced and revenues increased through automation in warehouses, factories and deliveries. Much of the supply chain can and will be automated by people that know the business and understand technology.

As automation increases there will be tough decisions concerning jobs and employees. There will also be jobs lost. As of 2018 the human element accounts for 30% of costs.

The global supply chain is critical to life and the economy. We are all reliant on the supply chain in some manner either as consumers or as part of a supply chain business.

The food we eat and the products we use are all dependent on supply chains. When supply chains work as they should they are a wonderful thing, but sometimes they do not and it can cause grief. The supply chain is prime for a digital transformation and it has already begun.

Delivery time expectations are compressing more every week. For example, over 100 million people subscribe to Amazon Prime. Same-day delivery is expanding and will be the expectation in many cities. Amazon Dash Buttons make it easy to press a button and your product is queued for delivery. Amazon is also building their last mile delivery network with a fleet of independently owned Amazon delivery vans. If you are not keeping up, you are 'prime' for disruption.

Being a technologist who works closely with the supply chain affords me the opportunity to see large scale computer - machine - human integration. There is so much low hanging opportunity for digital products and services.

Vetting suppliers is important to insure they are providing authentic products. Blockchain is already making headway into authenticating products. I described current and future use cases in the chapter, 'Blockchain - Transforming an Ancient Concept'.

Insuring supplier redundancy is another critical opportunity if you own a manufacturing or logistics company. The threat imposed by single points of failure must be eliminated or at least you must understand the risk. Many large companies have been impacted or put out of business by not planning ahead to reduce risks due to single points of failure. The Fukushima tsunami and subsequent reactor meltdown was one such event that impacted companies large and small. Major recent hurricanes such as Sandy, Maria and Irma put many companies out of business.

I suggest you map your supply chain from supplier to customer. Insure redundancy throughout and find ways to improve. Business Continuity Management and Asset Management software can automate much of the process. You want to understand all your assets and possibly threats that can take you down. Doing a Business Impact Analysis is critical to understand risk and opportunity. Failing to plan is planning to fail. If you are concerned about risk and business continuity I go into them in great depth, through experience, in my 2017 book, The Ultimate Business Continuity Success Guide. More information is on UltimateBusinessContinuity.com and Amazon.com.

JDA (https://jda.com) has been a leader in sophisticated supply chain software for a long time. Great software is critical, as customer demands can change in an instant due to trends, global events, weather, moves by the competition and more. These are the things you need in to consider. Remember, you can never get blindsided.

I like the way JDA has stayed on the cutting edge with their new Luminate product, rather than sitting back, being too comfortable and getting disrupted. Luminate runs on a cognitive, connected and open platform. It is a next-generation digital solution that can turn real-time data and insights into fast, profitable business decisions, which is exactly what we have been discussing in this book. It leverages AI, big data, advanced analytics and IoT devices. It can even take input from drones and other mobile vehicles. All of this leads to smarter, more agile supply chain transformations for greater results. They have a cool video on their site which shows some of the edge technologies they leverage.

Warehouses and Factories

I see automated warehouses and factories as magnificent examples of integration and innovation of software and connected hardware.

Before I became a director with a leading logistic company, I was a senior technology officer with a top ten global financial company. I came to the logistics industry with the preconceived notion that their software would be elementary and perhaps written in decades old Cobol, Fortran or BASIC 1.0. Wow, was I ever wrong!

Imagine the coordination it takes to inventory and then deliver millions of products a day. Imagine robotic machines that can automatically pick-and-pack products in warehouse racks six stories high. Imagine miles of fast moving conveyor lines converging in one stream with an endless number of products a mere inch or two apart. Imagine the sophistication of routing all these products to the proper trucks across scores of loading docks and finally reaching the destination of the customer in a timeframe that delights them. You cannot do that manually and part of my job it is to insure technology is always available and revenue keeps flowing.

As we discussed in the chapter, 'Robots and Drones - Creating Extraordinary Mobile Opportunities', Amazon uses thousands of little orange robots that scoot around their cavernous warehouses picking the right products from tens of millions of SKU's. The robots then bring back a basket of products to a human who puts them in a box and prepares the order for shipping. All that automation saves a great deal of time and money and generates loads of customer happiness!

Servus (https://www.servus.info/en/) offers an amazing high-rise optimum flow process case picker using very cool autonomous transport robots that ride elevated rails. The system has a capacity to pick thousands of products each shift and deliver them to conveyors that lead to delivery vehicles, all without any human effort.

Servus provides short feedback loops and fast error detection resulting in improved quality for processes. As a technologist, I find watching this type of device is even better than visiting Disney World. Servus has a 2-minute video at https://youtu.be/sAffLJfccmU or from their web site. If you do not think this device is cool, I would be shocked.

RFID (radio frequency identification) is a popular way to track, inventory and locate products of all kinds. I have experimented with both active and passive RFID sensors. The active ones emit a signal indicating where the object wearing the RFID tag is located. The passive RFID tags are tracked from central locations that bounce signals off the RFID sensors.

Depending on your use case, either or both can be good solutions for tracking products through the supply chain. Active RFID used to be too expensive for many scenarios, but as with all technology, the cost has dropped dramatically and the places it can be used has skyrocketed.

RFID guns are typically used to scan incoming and outgoing product in a warehouse. The guns may have some competition coming from products such as ProGlove (proglove.de).

ProGlove is an IoT device that has RFID capability attached to the back of a work glove. Benefits include productivity, worker satisfaction and improvement of ergonomics. The glove can help you save time on each pick. Any time you can cut seconds off a process that is repeated many thousands of times there can be significant ROI and possibly disruptive opportunities. The glove allows the wearer to have both hands free to pick, pack or do other tasks rather than having to dedicate a hand to holding an RFID glove.

I discovered ProGlove at techDayNY, a huge technology show that has launched numerous digital giants. Shows are a great way to discover cool new products. ProGlove has found success in Europe and in 2018 began offering product in the United States and Canada. Companies in Europe, such as John Deere and Lufthansa are already benefiting from ProGlove. Their web site has many such success stories and metrics.

In addition to RFID, in the chapter, 'Internet of Everything', we also discussed many different types of sensors that can provide significant return on investment in factories and warehouses. Sensors costing only a few cents integrated with IoT devices can provide you with information on heat, humidity, flooding, etc. These events unchecked, can cost companies hundreds of thousands of dollars during even one large scale event.

Better, Smarter and Faster Delivery

We all want to get our 'stuff' faster and cheaper. Autonomous delivery by ship, truck and drone is becoming a reality and soon will be common-place. Plus, instant delivery using 3D printing technology may eliminate the need to send certain physical products at all. A 3D printer digital plan could be sent to the customer and they can print out the product.

Maritime Shipping and Commercial Trucking are core to moving product through the supply chain. Whether it is crossing oceans or lakes, picking up product at ports or rail yards or delivering it to ware-houses or customers, safely and efficiently moving product is important.

KONGSBERG (https://www.km.kongsberg.com) is developing autonomous / unmanned / self-driving ship control systems using integrated sensor technology and automated collision avoidance. They hold the world's first contracts for commercial delivery of autonomous vessels.

One of their many autonomous ship projects is YARA Birkeland which they envision as the world's first fully electric and autonomous container ship, with zero emissions. KONGSBERG is responsible for all key enabling technologies including the sensors and integration required for remote and autonomous operations, in addition to the electric drive, battery and propulsion control systems.

YARA Birkeland will initially operate as a manned vessel, moving to remote operation in 2019 and is expected to be able to perform fully autonomous operations in 2020. There is a cool animated video of the ship in action on the KONGSBERG web site.

Oskar Levander, Rolls Royce SVP said on CNBC in April 2018 that autonomous shipping will reduce costs and improve safety. As with cars and trucks, most maritime accidents occur due to human error or fatigue. Autonomous boats and ships will be able to be built lighter and simpler. In

November 2017 Rolls Royce unveiled their Crystal Blue luxury yacht with advanced digital technology and hybrid propulsion. In 2020 we will see small autonomous commercial vessels such as tugboats. Eventually, large cargo ships will be self-sailing.

Do not be surprised if you see a self-driving truck on the highway in the next couple of years. In October 2016, Anheuser Busch delivered 50,000 cans of Budweiser with an autonomous semi-tractor trailer from Uber. The truck did all the highway driving and the driver just hopped in the driver seat for the last miles through the city. Police officers that saw it on the highway said that it was driving better than most trucks they see.

Waymo, Tesla and Uber are making strides toward getting autonomous trucks in production. They all have test vehicles on the road. Not only will this make delivery more efficient and less costly, but it will save thousands of lives. Approximately 4,000 people a year die from their vehicles being hit from behind by a truck due to human error.

Amazon is close to being able to deliver packages to your car's trunk. They are also exploring building their own delivery service using cars, trucks and drones (more info below) as well as leveraging assets such as their Whole Foods locations and University drop off sites.

Using AI, big data and predictive analytics retailers will be able to load products on delivery vehicles before you even order them. The retailer will examine patterns, find trends, mix it with environmental data streams and determine the likelihood you will be placing an order. That can compress delivery time to minutes! Sure, in the beginning they may needlessly load product on their vehicle, but over time the algorithms will get better, as they always do, and the accuracy and service will provide amazing delight.

You will also be able to issue one-time electronic keys to delivery people so they can drop off packages in your house, rather than leaving them outside where there is a risk they can be stolen or damaged during inclement weather. You will be able to monitor their movement through a Ring type camera device.

Self-Driving Trucks meet Life-Saving Electronic Log Devices

For many years, most professional truck drivers recorded their hours of service (HOS) in paper logbooks. These proved cumbersome, prone to error and open to possible falsification on the number of hours driven in a day. Drivers working 20-30+ straight hours led to safety issues due to fatigue and human error.

Commercial trucking fleets in the United States were mandated to have electronic log devices (ELD's) installed by December 17, 2017. These devices accurately capture driver hours with the goals of improving safety and reducing fatigue caused accidents. ELD's track a driver's hours of service electronically by integrating with the truck engine.

The HOS regulation limits truckers to driving no more than 11 hours a day within a 14-hour workday. Drivers must then be off duty for 10 consecutive hours. In Canada, truckers cannot drive more than 13 hours a day within a 16-hour workday. Drivers must then be off duty for eight consecutive hours.

To meet the demands of the new regulations, powerful digital solutions are available. For example, **EROAD** (http://www.eroad.com) offers many sophisticated hardware / software products for commercial truckers that cover ELD and tax compliance, safety and fleet management.

Their EROAD ELD is a user-friendly in-vehicle device that features an intuitive touchscreen and synchronizes with the engine to automatically record hours of service. They package their service with EROAD's secure web-based portal, which provides tracking, real-time notifications and reports.

When self-driving trucks are available, they may extend the miles traveled during a work-shift and still comply with regulations. Potentially the truck can partner with a human driver, as in the case of the Anheuser Busch delivery I described earlier. The truck can do the highway driving and the driver can do the 'last mile' and city driving. When not driving, the human driver could sleep. The total hours driven by the driver will be greatly reduced but the possible total mileage will increase due to automation. We shall see if that becomes a reality.

Autonomous or semi-autonomous trucks can also help when the human driver is behind the wheel. Technology can provide them with safety and efficiency information, much the way an airline pilot and aircraft partner.

For example, telematic systems, which monitors driver's skills and habits, as well as engine performance, are already a core technology used by commercial fleets. Telematic systems can produce data that will improve mileage, lower emissions, lower fuel costs, reduce vehicle wear-and-tear and most important, improve safety.

Element Fleet (https://www.elementfleet.com) and **Geotab** (https://www.geotab.com) are two companies with interesting telematic solutions.

Eventually, fully self-driving trucks will drive entire routes from beginning to end. There will still be the challenge of unloading trucks at delivery points along the routes but that can be done with a lower wage person and eventually a robot will do the work. Unfortunately, drivers will lose jobs. As of 2018 one of the biggest expenses for logistic companies is driver salaries. Hopefully, drivers can be retrained for other positions.

Commercial trucking can also benefit from situational awareness tools such as **RiskPulse** (RiskPulse.com) to monitor weather, traffic and road conditions. I use many situational awareness products for various use cases. One of the many strengths of RiskPulse is providing weather insight for logistic companies.

Using RiskPulse, you can map weather events to your delivery routes on a map. The cherry on the cupcake is mixing geo-coded latitude and longitude coordinates sent from your trucks into the soup and you can monitor their position in real-time as they enter and exit areas of risk. This insight can increase safety for the driver and the product, while reducing delivery delays and providing the customer with accurate delivery time-frames.

Drone Delivery:
PwC estimated the global market for commercial applications of drone technology will be valued at over $127 billion by 2020. There will be thousands of business opportunities and jobs created by this industry.

Drones already deliver crisis supplies and medication to areas impacted by disaster events. They will soon become important parts of the commercial supply chain. Proper safety regulations will be enacted and we will benefit from fast and safe drone deliveries.

As early as 2016, during a three-month trial run, DHL integrated their **Parcelcopter** (https://www.dpdhl.com/en/search.html?q=parcelcopter) into their delivery chain. It was a signal of things to come. People in the Bavarian community Reit am Winkl could use the automated service to send and receive packages. A total of 130 shipments were sent over eight kilometers – a trip that would take 30 minutes by car, but took the Parcelcopter only eight minutes. It was a resounding success!

Flirtey (flirtey.com) is carving out a niche in drone delivery. In 2016 they partnered with 7-11 and delivered food and a Slurpee to a resident in Reno Nevada. Flirtey has more recently announced a partnership to launch the first automated external defibrillator (AED) drone delivery service in the U.S.

In New Zealand, a man flew his drone to a driver pickup window with an order attached and flew the food back to a park as a surprise for his girlfriend. This was a controlled experiment in coordination with KFC staff. This is not something you should try.

UberEats (https://www.ubereats.com/en-US/) announced in May 2018 that they want to deliver really fast food - delivered by drone in 5 - 30 minutes. At the Uber Elevate Summit, Uber CEO Dara Khosrowshahi said, "Uber can't just be about cars, it has to be about mobility. It's my personal belief that a key to solving urban mobility is flying burgers, in any city. We need flying burgers."

Finally, to continue the fast food theme, in the opening scene of the blockbuster movie 'Ready Player One' a real drone delivered a Pizza Hut pizza to a resident in the 'stacks' in 2045. For many theater goers, it might not have been exciting, but for one digital technologist / futurist... I wanted to rewind that scene over and over!

Paul Misener, Vice President for Global Public Policy for Amazon.com appeared before a congressional hearing in June 2015 as part of his insights he shared, "Amazon Prime Air is a future service that will deliver packages of up to five pounds to customers in 30 minutes or less using small drones, also known as "unmanned aircraft systems" or "UAS."Flying beyond line of sight under 500 feet, and generally above 200 feet except for takeoff and landing, and weighing less than 55 pounds, Prime Air small UAS vehicles will take advantage of sophisticated "sense and avoid" technology, as well as a high degree of automation, to safely operate at distances of 10 miles or more, well beyond visual line of sight.

Not only do we think our customers will love this service, we believe it will benefit society more broadly. Once operational, Prime Air will increase the overall safety and efficiency of the current ground transportation system, by allowing people to skip the quick trip to the store or by reducing package deliveries by truck or car. For the same reasons, Prime Air will reduce buyers' environmental footprint: If a consumer wants a small item quickly, instead of driving to go shopping or causing delivery automobiles to come to her home or office, a small, electrically powered UAS vehicle will make the trip faster and more efficiently and cleanly."

If you are a consumer or your business relies on a supply chain, be prepared for many changes and improvements in the near future. I hope this chapter sets you on a path to researching and implementing technologies that can deliver new customer value.

Smart Sports, eSports and Wearables

Smart sports, eSports and wearables go beyond fun and games. They are three massive opportunities you must have on your radar from a business and career perspective. Each will be a billion dollar niche in the near future and will spawn many cottage industries.

In this chapter, we will learn about each of these niches. Even if you are not a manufacturer or currently in the sports or apparel businesses there will be many other opportunities to get a piece of the pie including: social and news platforms, communications, marketing, team ownership, IoT and sales. Be creative, be different and position yourself to provide value.

Smart Sports

Smart devices applied to sports can often mean the difference between winning and losing. Improving an inch on each stride can significantly improve a runner's time in a mid or long distance race. Athletes at any level and in any sport, can improve their games using smart devices. Athletes, including myself, pay for information and tools to attain that edge.

Fitness trackers have steadily grown in popularity and smart watch sales met with a bigger than anticipated user acceptance after a slow start and will increase at a rapid rate in the future.

I am a smart device success story. A few years ago, I took a devastating fall in a basketball game and made the gut-wrenching decision to 'retire' from the game I had enjoyed for many years. With zero experience, I took up racewalking and set my sights high. I embraced the challenge of having fun and winning a national racewalking championship. Key to my plan was to log and analyze my steps, time, pace and elevation using a fitness tracker for clues to improving my performance.

I became a **Fitbit** fanatic and currently have over 25,000,000 steps logged on their platform. Fitbit wisely allows users to form online teams so that they can compare steps with friends, family and colleagues and even experience some friendly competition. Like any true platform, you can earn badges and points. It was bittersweet when I received their highest distance badge in July 2018 called Pole-to-Pole. It indicates having walked 12,430 miles, the distance from the North Pole to the South Pole. I hope they create more badges for further distances.

Corporations can integrate into Fitbit analytics through an API and run contests with prizes to help employees stay healthy and have fun. There are other fine fitness trackers so do not feel you must use Fitbit.

As a serious athlete, I kept accurate charts and tried to steadily improve every day. I experimented with new counterintuitive techniques. For example, in racewalking small steps at a high turnover are way better than long lopping steps. Some racewalkers can go under 5 minutes a mile with tiny fast steps. Using the feedback from my smart tracking devices it was easy to test and find the techniques that worked best for me. I 'practiced with a purpose', never considered quitting and put one foot in front of the other - a lot.

Within 6 months I competed in the New York State Racewalking Championships and won the silver medal. Three months after that I won the United States Amateur Track and Field (USATF) National 10k Championship. Six months after that I took the silver in the USATF National 5k Racewalking Championship. I then applied the same digital data techniques to running and won many local 5ks over the next year.

The point is, data can become insight and help you improve performance in sports, business or any other endeavor. I would not have had the success and fun I achieved had I not used smart digital technology.

As an athlete, technologist and business leader, I get excited about the sampling of solutions and innovative ideas listed below. You will realize athletic manufacturers are sold on digital, which is great for us. As you read through these examples think of ways you can bring intelligence to everyday products for added value. It does not have to be sports products. It can be anything from a simple toothbrush to a more complex mobile phone. Everything is in scope:

As a long-time basketball coach and serious player, I can see the benefits of The Wilson® X Connected Basketball. It connects to Bluetooth via your phone or tablet. It has four playable modes programmed in that allows athletes to work on a variety of skills. You can challenge your friends and coaches can challenge their players. You can show stats and quantify your improvement.

Embedded in the smart ball are:
- Smart sensors that track makes and misses
- Sensors that connect via Bluetooth to the Wilson® X app to record your progress
- Technology to simulate game-time situations including crowd noise, clock countdowns and horns
- Battery power that stays strong for over 100,000 shot attempts
- There are four playable modes:
- Free Range: Tracks shooting percentage and distance from the hoop. The ultimate mode can be dangerous from anywhere on the floor
- Free Throw: Work on your routine and form while the app keeps track of your makes, misses and lifetime attempts. There are even a lot of pro's that could use this
- Buzzer Beater: Tracks 'made shots' that add more time to the clock. It is designed to get you ready for that game-winner at the horn
- Game Time: Allows you to be the star player against a virtual opponent in a real game simulation

The Adidas miCoach Smart Soccer Ball records strike point, speed, spin and trajectory when you kick the ball. It helps tune dead ball kicking technique or improve your ability to control and manipulate the soccer ball. I can see this helping many young players.

DribbleUp (http://dribbleup.com/) has taken soccer and basketball training to a smart new level. Their soccer and basketballs do not need embedded sensors or charging, as they use embedded optical marking for live ball tracking. I call that innovative thinking! The feel of the ball is like a regular ball. The app acts as a virtual trainer designed to improve your skills. You can watch action videos on their site.

Combining digital and physical, the Wilson® X Connected Official Football and mobile app provides five game modes while tracking each throw for real-time stats.
- Features five immersive game modes including QB Warm Up, Elimination Mode, Precision Mode, Game Time and Final Drive Mode
- Practice your throws with QB Warm Up
- Test and challenge your skills with Elimination Mode
- Throw your best with Precision Mode
- Play as or against any NFL® team in Game Time Mode
- Move the chains and find the end zone with Final Drive Mode
- Track each throw for comprehensive stats
- Offers a breakdown of the velocity spin rate, spiral efficiency, distance, and Wx rating
- Create your own avatar
- Post your scores and stats to see how you rank
- Link your social media account with the Wilson® X Football app to share

Greg Norman is a legendary professional golfer who has built a world-class brand. He is taking golf to the next level by applying the benefits of digital technology. In 2018, GN Media, an affiliate of the **Greg Norman Company** (GN) (http://www.shark.com/) announced EZLinks Golf (https://www.ezlinksgolf.com/) as a technology partner for Shark Experience. EZLinks worked with GN Media to develop "Shark Key" – a new product that enables a seamless, cashless experience from course to clubhouse that is central to the Shark Experience platform.

EZLinks delivers connectivity, content and customization to the course through a connected golf car experience. With the introduction of "Shark Key" to the Shark Experience platform, golf course owners and operators have a secure technology for driving increased revenue outside of the pro shop. As the product continues to evolve, the cashless system will provide a "member-for-a-day" experience, allowing golfers to purchase food and beverage and real-timehole-in-one gaming from their golf cart without ever taking out their wallet. "Shark Key" is compatible with all point-of-sale systems currently used by golf courses.

Greg Norman said, "The cashless concept of 'Shark Key' has shown success in many other industries such as hotels and amusement parks, and EZLinks helped enable us to bring that innovation to the golf course. Ultimately, we are providing golfers with the convenience they desire."

Golf is a very big business and there are many other ways golf is becoming digital such as with smart balls, clubs and improved digital distance finders. In the future, you will be able to play a round of golf with friends located anywhere in the world using augmented and virtual reality.

Tennis has seen major benefits from technology. **Cyclops**, which has been referred to as an 'electronic line judge', provides information on balls in play or out of bounds. It has been used for 45+ years.

Hawk-Eye (https://www.hawkeyeinnovations.com/sports/tennis) is a more recent electronic line judge that uses 6 video cameras. It is now used in tennis, cricket, badminton, hurling, rugby and volleyball. It visually tracks the trajectory of the ball and displays a profile of its statistically most likely path as a moving image.

In/Out (https://inout.tennis) is a cool idea. A portable very low cost ready-to-use line call device that is much more affordable compared to professional $50,000 systems. You attach it to a net post in 45 seconds and it delivers real time data that can improve your game. It also provides real time light and sound notification of line calls. It can even export data in cvs (comma separated values) format so you can integrate it with other programs. I have not tried the device yet, but the value and cost benefits make it seem like a potential digital winner.

Smart tennis rackets and sensors are very cool and valuable. **Babolay** (http://en.babolatplay.com/) is a leader and has been producing rackets that deliver metrics including spin, number of shots played and power. Rafael Nadal has trained with their Play racket.

Zepp (http://www.zepp.com) is an innovative company that sells pop-on, easy-to-use, sensors for smart baseball, golf, tennis and softball. It is very affordable and the included app delivers a lot of metrics to analyze and improve your play. This is another product in my near future or as a gift.

eSports

eSports has opened up a digital opportunity for new games, leagues and informational products. The physical sports ecosystem will provide you with clues to opportunities in the virtual world from a participation, business and career perspective. If you think eSports is only a game you are partially right, however it is also emerging as a huge business opportunity. Someday it will incorporate augmented and virtual reality.

Males under the age of 25 watch more eSports content than they do live sports. eSports is being considered as a future Olympic sport.

eSports competitions can sell-out major arenas in minutes. The top players can easily win over $1,000,000 in prize money and sponsorships are blooming. I enjoy watching eSports competitions on ESPN and other sports channels.

Some colleges have started giving scholarships to the best eSports players. Camps and schools, such as IMG in Florida have eSports training programs to improve skills and physical stamina necessary to compete on a high level.

Gamer Sensei (https://www.gamersensei.com/) developed a matchmaking platform for eSports athletes and coaches. Coaches who can improve a player's skills and scores is a thriving business. Coaches on the site must pass a rigorous test for playing and teaching ability.

Players select the game they want to improve in and the available coaches are displayed. When I visited Gamer Sensei there was a large list of coaches to choose from and hourly coaching prices ranging from $3 to $187.50.

Skillz (Skillz.com) is a major producer of eSports competitions and broadcasts millions of minutes of tournament action. They are a platform and much like other popular platforms they do not depend on owning physical assets. They bring people together and provide a service very much in demand.

TwitchPrime (https://www.twitch.tv/prime) - Amazon 'followed the money' and in 2014 acquired video-streaming service Twitch Interactive, which was formerly JustinTV for $970 million. TwitchPrime is included in Amazon Prime, which has over 100 million members. ESports broadcasts are a popular feature and Amazon's participation will help explode the growth of eSports.

I am confident eSports will rival physical sports in popularity and as a business in the future. There are already established sports teams buying into eSports league and team ownership. They will be able to do crossover marketing and further grow the eSports industry. Many careers and businesses will be built on eSports.

Smart Wearables

Ordering pizza from your sneakers? Pizza Hut was very creative during the 2018 NCAA Men's Basketball Final Four. Even as a digital expert and futurist I was not sure it was real when I watched a commercial showing someone ordering a pizza by simply pushing a button on the tongue of their sneaker. The technologist in me was compelled to immediately freeze the Final Four on my DVR and research what it was all about. The game could wait!

I learned it was real! Pizza Hut placed a blue-tooth enabled sensor in the sneaker which when pressed communicates with their mobile app. The app sends a message to Pizza Hut and uses your mobile device geo-positioning information for the delivery address.

Although some may look at pizza ordering buttons built into sneakers as a bit over the top, I think it is an interesting experiment by a big innovative company. It demonstrates that innovation and creativity is not limited to startups. I will bet some 'intrapreneurs' at Pizza Hut brainstormed the idea and it became a reality. To thrive and survive big companies need people like you that know digital and think outside the box.

I am sure Pizza Hut will analyze the metrics closely to see buttons are a commercially viable ordering path. Hey, if the popular Amazon Dash button is any indication, it may lead to all sorts of items we will be able to order from easy to reach button-type devices embedded in our clothing or perhaps in our bodies.

I mentioned fitness trackers at the top of this chapter that currently monitor steps, pace, speed, heart-rate, etc. As more monitoring capabilities are added to these devices going to your doctor every year for a 'point of time' checkup will be a thing of the past.

Unobtrusive fitness trackers and apps will privatively monitor your health. Much like the sensors that monitor General Electric jet engines to identify preventive maintenance opportunities before a disastrous breakdown, these wearable fitness/medical devices will securely monitor your heart, lungs, blood pressure, kidney and other vital organs and signals. When a predetermined threshold is crossed, actions will occur which can lead to warning you and your physician in real-time of an impending event. I call that a wonderful use of technology.

The U.S Olympic team that participated in the frigid temperatures in Korea during the Winter Games of 2018 wore Ralph Lauren produced jackets with embedded sensors and heaters with an 11-hour battery life to keep athletes at just the right temperature prior to competing and during the opening and closing ceremonies. This helped performance and the ability to score medals.

The limited-edition jackets sold out quickly but the technology will surely be mainstream in the near future. I do not like cold weather and would buy one in a minute, so count me in.

Now it is up to you to use your imagination. You can make any garment, sneaker, shoe (remember the Get Smart Shoe Phone), watch, bracelet, ring, necklace, glasses (think Google Glass), etc. smarter and deliver disrupting value. Even a sneaker that can order pizza!

Smart Digital Communications
Tools and Technologies that Work

Communication is core to humanity and your business and career success. New digital communication tools and technologies are shrinking our world by providing better, faster and cheaper ways for people and businesses to connect under any circumstance. Digital also makes one to one, one to many and many to many communications easy and flexible. For the first time in history, everyone can be connected. Digital eliminates boundaries, borders and geographic challenges.

Digital communication has disrupted the cost of staying in touch. In the industrial age, long-distance phone calls cost $5 per minute or more. Cost was based on distance, and in some cases, the country you were calling. I remember how we would rush to make our calls as short as possible.

Now, digital long-distance calls, conference calls, videos, webinars and podcasts are all low-cost or free. I rarely use a traditional desk telephone for voice calls anymore. I use Google Voice for most of my meetings and I sometimes mix in Skype for Business. We can round-down the cost of communicating to free and the voice quality is excellent.

Failing to communicate with customers or employees or communicating the wrong message is a sure way to get disrupted by a competitor that steps up and takes advantage. Communication can be a threat or an opportunity for your organization. In this chapter, we will learn how to make it an opportunity. There are so many new ways to communicate, it is mind boggling. There is no excuse not to do it well.

Throughout the book, we discuss social networks and you probably already know most of the big ones so I won't spend too much of your valuable time on them in this chapter. Twitter, Facebook, LinkedIn, Instagram and Snapchat are a few excellent ways to reach large audiences. These tools can ignite political uprisings or spread celebrity baby pictures across the globe in seconds. The power to freely reach and inform millions of people or highly targeted groups has disrupted industries including advertising, education, news media and sales.

I build many business relationships using social media. If you are not using social media yet, it is time to begin. The best advice I can give you is to give value to get value. If you help people, they will often return the favor. I discuss the Law of Reciprocity in the upcoming chapter, 'How to Leverage Content, the Currency of the Digital World'.

Below are other incredibly valuable communication tools you must have on your radar. I have used all of them with great success:

Digital Mass Notification Systems can be valuable in many ways. I have used them for life safety for thousands of employees and to drive revenue by serving customers in unexpected ways.

They enable you to send thousands of consistent messages in seconds using multiple different delivery paths, such as voice, email and text. You can optionally receive important responses recipients concerning their safety, security and ability to receive deliveries.

For example, during hurricane Sandy I sent out 38,000 voice calls, emails, text messages and push notifications all in less than five minutes to account for employee safety. In addition, if you have many customers you can delight them by communicating with them during a disruption at the push of a button. That can be a revenue generator and a competitive differentiator. They will not forget your thoughtfulness. You cannot do that level of communication using manual call trees!

Everbridge (everbridge.com), is a leading mass notification software vendor. They have robust notification capabilities and many additional important benefits., For example, they have developed an extensive safety app. One of the benefits is you can tap a panic button in their app and it will send an SOS message to a predefined person or group of people.

You can even activate the SOS features without having to tap the button. This can be important if you are in a dangerous situation and cannot initiate a call with your phone. For example, if you will be going into a dangerous area you can set a countdown timer prior to leaving and if you do not stop the timer before its set time, a notification will automatically be launched to any recipients you indicated.

The Everbridge app can also automatically track your location if you activate an auto check-in option. Every 500m you move, a notification is automatically sent and a series of these notifications can be displayed on a map. This can be important to security in case they need to find you or determine the direction in which you are traveling.

Everbridge can deliver many valuable and creative customer benefits. I find that when I bring Everbridge into a company, word spreads fast and leaders come to me asking if they can use the system for other applications to improve efficiency and service their clients better, faster and cheaper. I am also asked about using Everbridge with IoT devices and alarm systems. Often, the answer is, 'Yes, Everbridge can do it'. There are so many use cases when you think like an Intrapreneur.

Emergency Hotlines are essential to keep the public and your employees informed during a crisis or other important events. Communications can make or break an organization and there is no excuse for any deficiency in that area.

Emergency hotlines are typically comprised of a toll-free phone number and/or an Internet bulletin board. They are the perfect complement to your mass notification system. Emergency hotlines are a pull tool - people call in to access important information. Mass notification systems are push tools. You need both for a robust digital communications infrastructure.

Emergency hotlines can use voice recordings, text to speech or content on web pages (bulletin board feature). They can optionally allow employees to speak with operators or post responses, pictures or video. An emergency hotline must be able to accommodate many simultaneous callers. Do not use a service that will only support 10-15 callers if your organization has 3,000 people, customers and vendors. Whenever you implement a system, always allow room for growth. If you think you will need to accommodate 50 simultaneous callers, – then implement capacity for 100+.

Ring Central (ringcentral.com) is a great customizable tool I use to build highly functional digital emergency hotlines, customer service IVR's and other important applications. I share more details about my experience using Ring Central and its value to customers and employees in the upcoming chapter, 'Why it is ALWAYS About Your Customer'.

IoT Alerts enable you to push important information from alarm systems, industrial machinery, distributed weather stations and many other types of smart devices.

IoT alerts allow these machines to 'talk to us'. They can either have push alert capability as part of their default features or an API to enable other systems to connect and leverage the alert information. For example, I use a personal IoT weather station that allows me to use an API to access lightning and tornado alerts and automatically send a mass communication using my mass notification system. This enables me to use two world-class tools to maximize safety.

Satellite (Sat) phones enable you to make and receive voice calls from practically anywhere on earth. The phones are relatively costly but when you are in a situation where you must communicate, they can be life-saving.
On the lower end of satellite communications there are some really cool 'ping' tracking transceivers such as the Iridium RockBlock and Globalstar's Spot Trace, which I use and describe below.

Spot Trace (https://www.findmespot.ca/en/) is an inexpensive 2" x 3" device, smaller than a deck of playing cards. It integrates GPS, a satellite transceiver and a power supply in a compact casing. It provides one way communication from the device.

In my testing the Spot Trace pings can go through wood, plastic and glass, although it has difficulty penetrating metal. It bounces latitude–longitude coordinates off low-orbit satellites to mobile or desktop devices. Software programs can plot the coordinates on a map. I own a Spot Trace and have successfully tested it with a recent 21st Century Message in a Bottle project I was working on.

It is a great solution to monitor the movement of assets. It can be hidden under a truck, car, boat, etc., so the vehicles can be tracked in the event they are stolen. It is not intended to be carried on a person. It stays in sleep mode to conserve battery life and activates when it senses movement. Three AAA batteries can last months. If you attach solar cells it can work for years.

Iridium Rockblock (http://www.rock7mobile.com/products-rockblock) provides two-way satellite communication. It can transmit latitude – longitude and additional information such as temperature or salinity readings from the middle of the ocean or from outer space. It is two-way communications, as opposed to Spot Trace, so you can send it instructions from anywhere on earth.

Rockblock can be coupled with many types of IoT sensors and motors using an Arduino or Rasberry Pi microprocessor for added intelligence. Rockblock is more difficult to set-up than Spot Trace, but in certain use-cases it is the preferred solution.

Physical (hard) panic button
(https://www.alertus.com/alertusproducts/panicbutton) are tiny physical devices that you can hold in the palm of your hand and when you are in danger, you can press a button to send an alert. A panic device coupled with amass notification tool can be of life-saving value.

Zello (zello.com) is an easy-to-use walkie talkie push-to-talk (PTT) app. My family is hooked on Zello. It became very popular during major hurricanes in 2017. I used it to communicate with people in Texas and even during the devastating power outages in Puerto Rico I was able to get important information from people using Zello on the Island. I mentioned Zello on Facebook and LinkedIn and many people appreciated it and installed it. Zello has spread virally and has over 100 million users as of 2018.

Zello allows you to create private or public channels. It uses streaming audio which makes it faster than messaging apps that use recorded messages. If you try Zello, use it sparingly, as it can be a drain on your mobile device battery.

Facebook Messenger, WhatsApp, Viber and **Telegram** are other very popular mobile messaging tools. If you are chatting with people in China **WeChat** and **QQ** are the way to go.

Slack (slack.com) is a great way to communicate with teams. It includes built-in work-flow benefits. I have used it successfully on numerous projects.

Yammer (yammer.com) is becoming popular in organizations as a collaborative communications tool. It can help break down silos. I use it.

Virtual meeting rooms elevate conference calls and webinars to a more immersive level. They can reduce travel expenses, while fostering collaboration. Three interesting products are SoCoCo, Second Life and rumii. I have tested and used SoCoCo and Second Life and I am excited about trying rumii, which may usher in VR social meetings and conferences.

SoCoCo (sococo.com) is a business tool that enables you to design a realistic virtual environment with private offices and conference rooms. You can have public and private meetings. You can even shut the door to your virtual office to speak with someone privately.

Second Life (secondlife.com) is a virtual world unto itself. It was ahead of its time when it debuted in 1999. **Philip Rosedale**, the founder, is a brilliant technologist and VR thought-leader. If you ever have the opportunity to hear him speak, you should do so.

In Second Life, you are represented as a customizable avatar. You can meet with your team in buildings or office space. You can even create buildings from the ground up! Large companies have purchased choice real estate on Second Life for retail stores, banks and insurance companies. It even has its own currency, Linden dollars, that can be traded in the physical world.

For some companies, Second Life did not live up to commercial expectations, while others have realized the value. Some entrepreneurs claim to earn $100,000+ annually on Second Life. Second Life has approximately 1 million users.

In 2017, Linden Labs, the creator of Second Life, debuted their next generation product, Sansar (sansar.com). It is a social VR platform. The goal for Sansar is to provide a creative VR medium, making it easy for people to create, share, and sell their own social VR experiences. It is Linden Labs long-term goal to create the "YouTube of VR," as per Ebbe Altberg, the CEO of Linden Labs.

In May 2018, Sansar announced partnerships with several professional teams from the Overwatch League (OWL). Sansar will be allowing Esports fans and players to interact in a new way by utilizing VR. It is early in the game but you should keep an eye on their environment and explore ways to leverage it.

rumii (https://www.rumii.net) from Doghead Simulations (http://www.dogheadsimulations.com/) is extremely cool. It provides an immersive experience for remote teams and meeting participants. It gives you the feeling of actually being in the same room as other attendees. It has the potential to disrupt the way we work and socialize.

A question I often receive is, 'all of these Internet tools are great for communicating, but what if the Internet is not available? Can I still use them?" The answer is, Yes and No. Mesh networks may provide a solution to using apps without Internet connectivity.

Mesh networks use a series of nodes (devices) to pass information upstream and downstream without depending on an Internet Service Provider. Instead a simple Internet Gateway Provider (IGP), who could be anywhere on the meshnet, can enable connections. Meshnets may eventually provide a greater level of communications resilience than we currently enjoy.

In the chapter, 'Smart Supply Chain - Our Chain of Life', I mentioned **CattleWatch** which utilizes solar-powered 'smart collars' that allow a herd of cattle to form a wireless mesh network. Ranchers can remotely manage their herd through connectivity via Iridium satellites. Industrial grade TLI Series rechargeable Li-ion batteries enable the 'smart collars' to be lighter and more compact, and thus more comfortable for cows to wear. The batteries are from Tadiron Batteries, NY.

FireChat (https://www.opengarden.com/firechat.html) is another example of mesh net technology in action. It was developed by Open Garden, a leading free messaging app for public and private communications that works even without Internet access or cellular data. Built with the **MeshKit SDK**, it connects smartphones to one another and lets users share and forward data even without the Internet. Data flows directly between smartphones (iOS and Android). Each node essentially becomes a router. As mesh networks grow they become more resilient and self-healing.

FireChat has a wonderful feature named 'FireChats' that allows 10,000 people to gather anonymously in a group chat. You do need to find local nodes that are relatively short distances apart so you can be part of the network. FireChat is continuously improving so read the latest requirements on their web site or in my special report.

Here are 4 example use cases from the FireChat web site as of August 2018:

- MeshKit Developers can reach users who are not connected to the Internet because they don't have a data plan for example. Engagement and sessions within your app are increased with a new form of device-to-device connectivity. Content delivery costs are reduced as data traveling over the mesh is free. MeshKit can also be used to distribute app updates to users over the mesh (Android only).
- Media companies can share and distribute content to users who are not online. They can source real-time news from users in the field, even when users or reporters are not connected. Reports will bounce back through the mesh until an Internet connected node is reached and then delivered to you anywhere.
- Telecom operators can provide customers with a new form of wireless connectivity that works when others don't. MeshKit provides a new way to manage increasing demand for mobile data.
- Government can send out early warning and recovery advisories before, during and after a natural disaster. They can reach more people regardless of cellular infrastructure or damage. Collecting citizen information can increase community resilience.

Digital communications tools and technologies are improving at an exponential rate of speed. There are many incredible new technologies and devices I am exploring. I will continue to share my new findings in the newsletter.

Step 4

Applying Process for Customer, Business & Career Success

In this section, we will put everything you learned together. You will learn best practices to build a digital program. This section is high on practical experience and low on theory.

I share hands-on experiences, tips and techniques that worked well for me. These include delivering enterprise solutions and entrepreneurial products and services.

I also share what did not work. Often it is these 'lessons learned' that provide the most value, on the condition that you learn from them.

You will benefit from stories and experiences of other successful digital entrepreneurs and executives. Apply their insight and you will be on the fast track to digital transformation success.

For your convenience, links to all companies and products discussed in this section plus new golden digital nuggets are listed on DigitalSuccessBook.com/links.

Why It is ALWAYS About Your Customer

Here are 11 'Customer Commandments' I live by:
1. Thou shall treat your customer as king and queen
2. Thou shall understand your customer
3. Thou shall communicate with your customer
4. Thou shall be interactive with your customer
5. Thou shall respect your customer
6. Thou shall deliver extreme value to your customer
7. Thou shall actively listen to your customer
8. Thou shall under promise your customer
9. Thou shall over deliver your customer
10. Thou shall always keep promises to your customer
11. Thou shall do everything in your power to make your customer successful!

If you understand your customers' requirements, alleviate their pain and exceed their expectations, you will succeed. Think about Apple, Amazon, Adobe, Google and Tesla. Then think about Motor Vehicles and the Post Office. Ouch. The first five are laser focused on their customer's needs and the last two perhaps not as much.

When you read the digital success stories in this book you will see how each innovator values their customers above all else. They pay attention to detail and create value at every turn. I believe this is a surefire way to success.

Everyone experiences customer delight in different ways. It is critical that you figure out what your customer values most. What makes them happy. For example, I love going to Starbucks. I look for excuses to visit one of their storefronts and have the privilege of spending hundreds of dollars a year on their coffee. But it is not just the coffee. There are other places I can visit if it were just the coffee. In fact, I do not even have to leave the comfort of my home, get in my car and travel to get a cup of coffee.

Starbucks delighting me must be more than the coffee and it is. I enjoy the product, the music, the environment, the social aspects and the newspapers. In fact, one of the apps I included in this book I learned of during a conversation with a friend at Starbucks. It is an overall great experience.

When my son comes to visit, it is a ritual for us to take our laptops to Starbucks. He codes and I write. We often do not even speak to each other the whole time except to say, "it' time to go". Sometimes we even text each other across the table. The clicking of our keyboards and hanging out together makes it a delightful experience.

Apple, Amazon, Adobe, Tesla, Starbucks, Disney...you know them all. They do it right and so do a lot of other digital companies you never hear much about that I will refer to in the book.

Tip - Work Backwards to Create Customer Value
One of my favorite techniques for providing extreme value is to work backwards from the end to the beginning. It works for me every time. You heard me correctly, work backwards and I do not mean sitting backwards while you work.

By the way, I thought I was unique in working backwards and then I learned that Jeff Bezos lives by the same technique. He is obsessed with understanding the value of the end result and then devising the solution to get there.

We will discuss the importance of developing a plan to get to the end result in the next chapter but before you can develop a step-by-step plan to get to value, you must understand where you need to go and what the end result looks like. You need a value target. Sure, using an agile project methodology the target might shift somewhat, which is fine and to be expected if you are staying attuned to your customers' needs.

Too often, I have seen problem solving approached in the opposite way. A solution is built without understanding the needs and specifications of the customer. It usually results in chaos, expense, failure and finger pointing on both sides. I believe that is why most digital transformation and software development projects fail. I have seen promising beyond startup enterprise companies destroy their businesses by not listening.

Understand How Things Get Done in the Real World
One of the best ways to know your customers, whether external or internal, is to spend time interacting with them. If you watch how they use a product, service or system and ask questions they will open up to you on how you can

improve products and processes. You can take their valuable feedback, make improvements, automate and possibly digitally transform and/or disrupt the marketplace.

Interacting with customers allows you to understand first-hand the customer journey. Each touch-point can be of value to you. It may start with how customers find your product or service and lead to how they order it, pay for it, how you source it, inventory it and pick, pack and delivered it. But do not stop there. Understand how they use it and how they share their great product experience with their network of friends, family and colleagues.

Every time you interact with people in a positive manner you are building social capital and strengthening your network. In the digital age, social capital is valuable currency. Think about platforms like LinkedIn, Snapchat, Pinterest and Facebook. They are built on social capital.

Customers will be so impressed that you are devoting time and effort to understanding their needs and that you value their input and are not assuming you know best. People do not like things forced on them especially when the solution is ridiculously off-base. I see it happen way to often with vendors and consultants.

If you have ever watched Undercover Boss, you know the CEO receives great actionable input and insight from employee interactions. Sometimes the interactions have such happy endings there are tears of joy.

In my opinion you cannot do innovation or a digital transformation sitting in your office with the door closed. Reading about it is important but seeing it done up-close is completely different and far better. I make it a habit of interacting with customers, riding with delivery drivers and salespeople and hanging out with warehouse employees. I learn something and discover opportunities to help them every day by doing so.

Customer Personas:

Customer personas can be useful to understand your customer. Daniel Kendrick wrote a nice concise article on digitalgov.gov:

"Personas are a tool organizations can use to learn more about their users. They are used to learn as much as you can about end users to better the product or service you provide. If you can think as a user during the design and development of a product or service, this will help greatly in creating something that satiates the users' needs.

Personas are descriptions that give you an understanding of your users and how they use your product or service. Personas can take on a variety of forms depending on your organization's needs. But, there are a few basic questions and ideas that every persona should include to ensure that the tool is useful. Every persona answers the questions listed below:
- Who is your audience?
- What is their background?
- What experiences have they had?
- What are they looking to accomplish?
- What are some challenges they face?
- What are some potential ways to address those challenges?

As you begin asking these basic questions you develop custom ones that fit your industry. The goal is to get a complete understanding of your users."

I agree with Daniel. For example, one of my most important internal customers is upper management. These are the executives that sponsor and fund our digital program. After building a customer persona through formal and informal interviews, focus groups and surveys I could build a map of their wants and needs. That enabled me to deliver many customer delighting benefits such as:
- Improving internal and external processes
- Eliminating redundancy
- Reducing expenses
- Growing revenue
- Reducing risk
- Improving data quality
- Mapping of our internal and external supply chain upstream and downstream
- Identifying disruptors entering our space
- Identifying disruptive business strategies our company could leverage to generate revenue and own our industry
- Identifying and mitigating cyber-risks that kept management up at night

When analyzing the wants-and-needs of external customers, I use customer persona questions along with interviews, focus groups, surveys, digital market research and social media mining tools.

I am so proud that 100% of the digital projects I participated in have provided value were successful. I attribute it to great teammates and customer partnerships.

Better Customer Service the Digital Way:
Every company spends time and effort trying to build a strong customer. Unfortunately, the same companies often neglect current customers.

In a world where great customer service is the exception, not the rule, you can easily differentiate business and your career by providing great customer service.

I am dismayed when I must wait 20 minutes to speak with a customer rep or when no phone number is listed on a company web site. If I send an email and do not get a response within 8 hours, it is disappointing and I look elsewhere.

Zappos (zappos.com)is a great example of a company that makes customer service a differentiator. They have gotten a lot of traction out of superior customer service. Tony Hsieh the founder and CEO even wrote a book, 'Delivering-Happiness-Profits-Passion-Purpose', that I enjoyed. It tells the story of the trials and travails of building a startup and describes the Zappos focus on customer happiness. Tony did it right and Amazon acquired them for $850 million.

New digital tools can help improve customer service without adding significant cost. Chatbots with integrated artificial intelligence can answer many more questions than a human and are much more efficient. They learn as they go along. In the chapter, 'Artificial Intelligence - The Future is Here and It is Bright', I name some chatbots that are surprisingly inexpensive. People are getting used to chatbots and most like using them.

LiveChat is a leader in chatbot technology. They have 24,000+ companies in over 150 countries using their product to help improve the customer experience. They have an impressive client base spanning enterprise and startup companies. I created a special LiveChat resource page with information on the benefits of their chatbot. I also include a no-risk 30 day free trial offer so you can kick the tires (as of August 2018). It is always important to try technology to make sure it provides value.

Web sites should have phone numbers prominently placed so you do not have to hunt for them. Call-hold-time should be held to a minimum. If you must reroute to outsourced agents, make sure they understand your product and act professional. Background noise should be at a minimum so it does not seem like the call-agents are in a boiler room taking calls for 100 other companies.

Figure out the lifetime value of your customers and it will pay to spend a little more per call on knowledgeable agents.

Build a knowledgebase and list of FAQ's consisting of questions and answers and keep it up to date. Every call that comes in should be an opportunity to improve your service for future callers.

Keep metrics on length of calls and to see if questions were answered to the full satisfaction of your customers. Strive to exceed their expectations so it becomes a selling point for your company.

Ring Central (ringcentral.com), which I briefly discussed in the chapter, 'Smart Digital Communications - Tools & Techniques', is one of my favorite integrated communication tools. It can help you create a valuable and differentiating customer service experience.

It is easy to use and reasonably priced. You can easily create an interactive voice response (IVR) menu system that enables callers to make prompt selections from their touch keypad to access the information they need. You can also integrate hunt groups to locate the best reps to efficiently answer customer inquiries.

In less than 2 days I was able to build a customer delighting IVR system for a major corporation. Callers simply call a toll-free easy to remember vanity number that Ring Central secured for us. They then tap the 2-letter abbreviation for their state to get the information they needed. Callers do not have to waste time with needless layers of wordy menus to get the information they need. The system produces more customer connects and sales. Customers and management love the system.

Ring Central has a fully functional IOS and Android app. Using the app, the entire system can be administered remotely.

Ring Central is great software. I am proud to endorse them and be able to offer you a free trial. You can read more about my success with Ring Central and a no-risk free trial at http://DigitalSuccessBook.com/RingCentral.

In summary, customer service is a touchpoint that should be used to foster goodwill and create customer happiness. When I receive unexpected superior customer support - which I sometimes do - I ask the agent for their name and the email address of her or his supervisor. I immediately email the supervisor a thank you message and ask that they please let the agent know how much I appreciated her or his patience and knowledge.

I also, tell my family, friends and business associates about my great customer service experience and it often leads to them becoming customers.

It is far easier to keep your current customers than to get new ones. With call-center software and AI chatbots there is no excuse to provide anything but great service.

Opportunities and Success in Industrial-Age Companies

I am sure after reading the previous sections of the book on mindset, the power of emerging technologies and how they are changing our world that you want to apply the concepts, tools and technologies as soon as possible. But it is important that the digital transformation you implement aligns with the vision and capabilities of your organization to best serve your customers.

If you work for an industrial age company and are responsible for the success of your company's transformation to the digital age, you will have special challenges that often do not exist in a born digital age startup. You will also have incredible opportunity to improve older products and services. I know, as I have built digital for corporations and startups. I have been in your shoes and learned how to find many digital opportunities. I also learned how to overcome every challenge.

Most digital projects in mid and large industrial age organizations fail. Yes, over 50% of projects some costing millions of dollars, fail. There are many reasons for this high rate of failure. We will focus on three main causes of failed projects and how to overcome them:
- Lack of a clear vision and goals
- Poor communications within a team and with stakeholders
- Fear and resistance to change

I have led over 100 digital projects in my career (so far) and every one succeeded, although at times it was not easy. In this chapter I will share what has worked for me in every project I led. My goal is to save you time, energy, possible frustration and hopefully help you build a superstar digital program.

Unless you are the CEO or founder of the company, one of your first action items should be to meet with upper management (leadership) as early as possible. That will empower you to build the best and most on-target digital program possible. By the way, if you are the CEO or founder make sure the person or people directing your digital program read this chapter and hopefully the entire book.

If upper management actively endorses your digital efforts, you will get cooperation from the entire organization including process owners and their direct reports. You will be digital gold!

Begin with a high-level discussion to level-set the scope and goals of the digital transformation program. Will the scope be only external customer facing or will it include transforming internal processes as well? I strongly recommend you both in scope. There is often a great deal of low-hanging fruit digitizing internal processes but sadly they are often overlooked.

I do not think your digital projects will be a hard sell to management, as digital is mainstream news and can provide great return on investment. Many leaders will understand the importance of making use of creative thinking, innovation and new technology to create customer value. Hopefully they realize every business, including yours, must be in part a technology business to prosper and survive.

Emphasize to management that disruptors can steal your customers and potentially destroy your business. Share some disruption horror stories. You could use ones they are familiar with such as Netflix - Blockbuster, Amazon - Sears, Ericsson - Apple, Lyft or Uber - the taxi industry, etc. Also, research and provide examples of lesser known ones that map directly to your industry. Those will hit closer to home and will have a big impact on management. The more they understand the pain the more responsive they will be. Make it all about them.

Communicate your willingness to go 'above and beyond'. Demonstrate that you have initiative and want to help in any way you can. Management usually welcomes that type of team-oriented creative thinking and will understand the value that can come from your efforts.

On the other hand, if upper management does not understand the true value you and a digital transformation will bring to your organization they will not actively support you and it becomes more difficult to succeed, although not impossible in my experience.

I have encountered management that just does not get it and worked on my own time to develop digital value. Some of it I did in stealth mode beyond work hours as an 'unofficial passion project'. Eventually management realized the value when I presented prototypes and demos of what I had developed. They appreciated my dedication and innovative thinking and became great proponents of the digital program initiatives.

When you have the support of management results will be wider reaching and digital transformation will happen faster. It will make your life much easier as well. If you cannot get their support don't give up. It is only a 'speed bump'. Be positive, follow your passion and just get it done!

It is important to learn upper management's goals and expectations for the digital program. When you meet with them listen, notice their body language and verbal inflections. Read between the lines. Small details convey a lot of information.

One of my favorite questions is, 'what keeps you up at night?' That question has been worth its weight in gold to me in all facets of business including digital transformation, IT and business continuity. I always learn a lot from their response which is often, 'being blindsided and losing customers to disruptors'. It makes sense. When customers are lost and revenue suffers it is upper management that is held accountable.

Not having critical digital technology available is also a top response of management and I agree it should be. When your organization relies on digital, technology must work. If your digital product is not available customers and employees will suffer. I have seen technology failures destroy businesses. To mitigate the risk of technology not being available it is important to create digital redundancy and backup for internal and external facing systems. You would be shocked at how many large, supposedly tech savvy, organizations have 'been down' for unacceptable extended periods of time due to the lack of planning and simple safeguards.

Tip - Actively listen, listen and listen some more - absorb as many leadership pain-points and digital goals as possible. During meetings with customers, upper management, process owners or line workers, if you are speaking more than they are then you have a problem. Dial it back. You will be happy you did. You never learn anything while you are speaking. The old saying 'we have two ears and one mouth for a reason' is so true. Listening is a valuable skill most people do not have. Devote time to become a great listener. It will have a big impact on your career.

Do not let anything slip through the cracks. Maintain eye-contact. As soon as possible update your checklist, spreadsheet or database with takeaways from the meetings. Soon you will have an ocean of information flowing through your desk. Staying organized is a must. Do not let tasks slip through the cracks or you will get bitten in the end.

Tip – Communication is critical - schedule digital program steering committee meetings with upper management regularly over the life of your program. Most projects, digital or otherwise, fail due to a lack of communications. If all meetings are not required, it is ok to cancel. Management is busy so meet when it is important but cancel when you or they have nothing on the agenda. They will respect that.

Tip - Break down silos - if you are in a medium to large industrial age organization you will likely be working with many different department leaders that report directly or indirectly to upper management. Silos may be ingrained in the culture of the organization. Silos are more prevalent in industrial era organizations than digitally born ones. These silos MUST be broken down to improve communication. Developing a successful digital transformation program means data and information is flowing vertically and horizontally across an organization which will create value.

Tip - It is 'all about them'. Your elevator pitch should focus on how the digital program will benefit the exact audience you are speaking with. People are often afraid of change. They may be concerned with losing their job or may be embarrassed by not knowing how to do things with technology. These are understandable concerns.

Communicate the positive business reasons for implementing a digital transformation program and the negative aspects of not having one. For example, a digital program can make the organization a leader in the industry, gain new customers, retain current customers, derive new revenue and reduce expenses.

Then take the reasons to a level that really hits home at a logical and emotional level - keeping the company in business, maintaining job security, having the ability to pay their bills and the prestige of working for a leading company. If you deliver a powerful 'about them' message, managers will dedicate time to helping you build the digital program.

Tip – You really need a meal - I have led many different types of enterprise-wide cross functional projects, including digital transformation, business continuity, enterprise software development and implementation, etc. In my experience, a great way to get full support from all levels of management is if upper management hosts a lunch for you your first week of the program. Everyone loves a good meal.

The meal meeting will clearly demonstrate upper management's support and buy-in for you and the digital transformation program. To kick off the meal they should speak to the attendees about the company's commitment to becoming a digital leader, changing the culture, breaking down silos and becoming innovators and disruptors. They should emphasize it will be for everyone's benefit and job security and everyone must participate in making the digital initiative a big success.

Interacting with people that will be critical to the success of the program in a somewhat social scenario will further break down barriers. You might be surprised how infrequently these managers speak with each other. The benefits in getting everyone together and on the same page can be astounding! It will be beneficial in breaking down silos.

Meetings and time commitments from middle-managers and line workers will fall into place when they realize upper management is committed to building a vibrant digital culture. This will enable you to meet your project deliverable dates and bring products to fruition on time or early. There is nothing worse than missed deliverables and a 'forever' project.

In one visionary company, upper management hosted a kickoff lunch for me to meet all the middle managers that would be instrumental to the success of my program, while one of my counterparts in another region of the country met with the middle managers 'cold' (no upper management hosted lunch or even email prior to middle management meetings). Coming into the meetings 'warm' made my job easier and very effective. People knew I was on-board to help them 'get digital' and never get disrupted. My counterpart, on the other hand, was seen by the middle managers as an efficiency expert. People were avoiding him at all costs. Perception is everything! It is hard to change first impressions. Unfortunately, my teammate left the company within a year, while I succeeded.

Finding Hidden Opportunities for Digital Transformation
Let management know that as you build the digital program and examine how things are currently being done, you will gain a unique holistic understanding of how your company works from end-to-end. This can lead to new external customer facing opportunities and internal ways to improve processes, generate revenue and/or reduce expenses.

A great way to find hidden opportunities is to spend time with managers, office workers, warehouse and factory workers, delivery drivers and anyone else that that does the actual work. See how they do their job and take detailed notes. Do not only interview managers. Often, they are removed from the actual way processes and tasks are done. Also, interact with customers to see how they are using your products and make suggestions on improving them.

I guaranty the time you spend with these people will be worthwhile. Not only will it help you develop your program but it shows people you care about them. Thank them for their time and leave your business card with them. Let them know your door is always open to them and you value their suggestions. Remember, in a company digital is a team game and everyone must participate.

Here are three possible internal digital transformation opportunities you may identify in your organization. I consistently find these and many other 'low-hanging fruit' opportunities in companies of all sizes:

- Departments hoarding information in local spreadsheets which devalues the data, is an opportunity to streamline the process and increase

revenue with better cross process analytics. This reduces costs associated with input errors and avoids time being wasted inputting the same information. Think gold copy for success!
- Commercial truck delivery drivers using paper forms to log hours worked. That is no longer allowed as you learned in the chapter, 'The Smart Supply Chain - Our Chain of Life'.
- Salespeople inputting orders on paper where it could be migrated to a form on a tablet which instantly transmits the orders to the central database. This will reduce errors and compress the delivery time to the customer. That can mean product differentiation and revenue.

Why You ALWAYS Need a Plan

"A goal without a plan is just a wish."
Antoine de Saint Exupéry

"Plan your work and work your plan."
Napolean Hill

"Always plan ahead. It was not raining when Noah built his arc."
Richard Cushing

"Planning in all walks of life is your key to success."
Marty Fox

Success does not happen by accident. Almost every successful person including me has stated she or he had a plan.
    In this chapter, I share planning and execution techniques and tips that work for me every time in every facet of my career. I led digital programs, data center migrations, developed and implemented new digital technology tools. wrote 13 business/technology books and won a national athletic championship. In each case my success was very much dependent on developing and executing my plan.
    In the preceding chapter, you aligned yourself with managements vision and goals. You build relationships throughout your organization. You discovered how customers use your products and opportunities for new products. You analyzed how things are currently being done in your organization. I am sure many digital opportunities and projects were identified and prioritized because of your efforts. Now is the time to begin delivering the vision and delighting the world.
    Digital transformation, disruption, innovation and delivering customer value is a process comprised of multiple smaller projects.
To complete a successful project, you always need a plan:. -
- Projects always need beginning and end dates. 'To be determined' (TBD) or approximately or 'whenever', are not acceptable 'dates'.

- Tasks that make up a project always need deliverable dates.

    Your plan does not have to be long and complex. In fact, a simple streamlined plan that gets you efficiently to your goal is better than an unnecessarily complex plan.

    We discussed the value of working backwards in the chapter, 'Why It is Always About Your Customer. That will enable you to identify customer wants and needs and develop high-level ideas on what it will take to bring value to customers. Now you must create a step-by-step project plan that will get you there on time and every time.

    Verbal plans do not work. Without a written plan and deliverable dates, you will probably experience scope-creep and perhaps a 'forever project' that never seems to get completed. I see way too many of those in corporations. If deadlines are missed it will get ugly for you and your organization.

    Define the scope of the project early and have the deliverables signed off on and keep the signed off version for your records. If the scope of the project is a moving target once the development starts you will likely run into dreaded 'scope creep'. I have seen scope creep first-hand from both sides of the table, when I was in IT and as a user. Scope creep is bad. Scope creep can turn into 'forever projects'. I do not care if the development is being done in agile, waterfall or any other software development process. Scope creep can occur regardless of the process so do not be fooled thinking it only applies to in-house developed projects. It can happen in any project.

If your goals are big, you will surely have challenges along the way. Embrace the challenges as they appear. If people tell you it cannot be done, prove them wrong. Demonstrate the digital technologist and athlete's mentality. Most people give up on the two-yard line. Just find a way, and power across the goal line.

    When implementing simple projects, you can use a spreadsheet or simple database project planning software.

    When implementing more complex projects you probably will want to use a dedicated project planning tool, such as Basecamp, Trello or Microsoft Project. These help you accurately estimate dates, determine resources and define tasks. You can also link tasks to milestones and create dependencies, where completing a later task depends on the completion of a previous task(s).

Small steps lead to big success:

People ask me all the time how I authored twelve successful technology-business books, published over 100 digital software solutions for corporations, developed entrepreneurial businesses and won a national racewalking 10k title within one year of learning the sport.

These are challenging projects that can seem overwhelming. They cannot be done in a day or with one attempt.

One of my major success secrets is breaking every project into small steps. Every project or journey, no matter how complex, can be conquered with a series of small steps. Small steps are powerful, fun and they work. Small steps:

- Provide you with a consistent sense of accomplishment as you complete even the tiniest steps
- Are granular and modular so they can easily be reused in other sections of the same plan or to save time when building other plans
- Quickly add up to significant progress
- Are more fun than large overwhelming tasks
- Work every time!

For example, writing and publishing 'Digital Transformation Success Secrets' has been one of the most challenging projects of my career. There was a great deal of information I wanted to share with you and I had limited time beyond my day job to create it. I systematically completed my ambitious book project using small steps. My project plan included tasks for researching, writing, publishing, marketing and sales. Each section was sliced into tiny tasks with deliverable dates. Small detailed tasks insure you do not miss anything important and it keeps the momentum going with consistent wins.

Deliverable dates can make or break your career:

It is important to define deliverable dates for each task that makes up your digital project plan. Without deliverable dates and deadlines, projects can go on forever. which is not a good thing. I have seen many 'forever projects' in my career. As a former large scale project manager, I cannot stress enough that every project and the tasks that make up a project must have start and end dates.

Achieving your deliverable dates on time is great. It means you have delivered value on or before schedule. Perhaps, you exceeded expectations. This can result in positive feedback from management and customers. Hopefully, it can also result in more money in your pocket around bonus and raise time.

Missing deliverable dates, on the other hand, is very bad for you and your company

Unfortunately, it is all too common in my experience. Projects often get kicked down the road as priorities change. In the digital world, delay can be devastating. Many failed projects begin with missed deliverable dates. Everyone must be accountable for meeting their deliverable dates.

In my previous book, I tell the story about missing one deliverable date due to constraints beyond my control. Even though I 'carried a department' missing that date cost me my annual bonus. When my direct manager told me I was not getting a bonus for the first time in my career, she had tears in her eyes. She knew I deserved one but upper management had made the decision. Lesson learned. My advice is, if there are any constraints that will cause missed deliverable dates, document and communicate them as early as possible.

Here are some tips that may help you estimate realistic project load, deliverable dates and deadlines. These originate from first-hand experience:

True Story – It is mid-November. Everyone is in a great mood, as it is holiday and bonus time. It is my favorite time of the year. It is also time to have an off-site meeting to plan for next year and have some fun.

On the agenda is a ridiculously long list of projects we will complete the following year – all 27 new systems (true story when I was in IT). Well, we have 5 people in the department and we are going to build and implement 27 new systems? Even the VP who is presenting this at our fancy-resort off-site meeting was apologizing for these upper management, you have got to be kidding me marching orders. Needless to say, many of the projects never were completed or even started. Morale of this story - do not over-estimate what you/we can get done because you will be held accountable.

Maybe you are the new digital technology director that wants to impress management or perhaps you are the CEO that wants to impress shareholders. So, you make promises that you will digitize your physical product in 6 months and it will go viral within three months or depending on the heat you are facing, maybe three weeks. Again, you are setting yourself up for disaster.

My advice is to not get caught up in the moment. Do not over-estimate. It is better to be honest and to set realistic expectations. You may want to set the bar a 'teensy-weensy' bit low and then work your butt off to over-deliver in a big way! Under promise and over deliver.

Do thorough research. Perhaps you have a project to implement a new HR or mass notification system. You read a very positive report, by one of the big advisory firms on a possible vendor solution. Cool. You set a production delivery date of 3months out. You put it in writing to management that you will get it done. Hey, 3 months seems like a long time, right? A quarter of a year – wow – that is a long way off – it won't come so soon – 90 days – a couple thousand hours – what could go wrong?

Well everything can go wrong! Time flies! Do your research. Talk to other buyers that have implemented the solution. If the project is software focused read the chapter on buying software and never getting burned later in the book.

Make sure your IT staff has the resources to support your digital project. Make sure they have the skills or have them take training. You also have the option to outsource portions of the project.

Make sure everyone is doing their job and keeping up with every aspect of the project. Project plan reports can help you identify any delays. Otherwise, it quickly can turn into finger pointing between internal IT and the vendor, with you in the middle. You DO NOT want to be there, I promise you. For those of you that have been there you are nodding your head in agreement right about now and if you are currently in that uncomfortable situation you are not smiling. Am I right?

As I mentioned earlier break each step and milestone down into detailed tasks. For example, when implementing an intelligent mass notification software tool, a major deliverable is getting accurate HR employee contact data into the system and building a process to insure it is kept up-to-date and accurate. That one step alone might entail multiple meetings with HR, Legal and IT. Then there is the long process of updating the employee information. Dealing with the integrity of employee data is critical to the success of the project. Unfortunately, employee contact data quality may never have been addressed prior to your project. It can take months just to get that right. The point is, you must factor in and document all facets of the project in your deliverable time-line.

Meet frequently with your teammates to discuss progress and unforeseen challenges. Sometimes you must adjust your project plan during the development life-cycle. When I was close to completion of this book I wanted to share important late breaking information with you before it went to print. I was easily able to modify my plan and add the content. Modifications are to be expected to create maximum customer delight.

When you reach a significant milestone, you can have a little party. When you successfully complete the project, you can hold a big lunch or dinner party at a nice restaurant.

Also, give team members some swag. If you are working with digital developers they love t-shirts, stickers and stress balls. Customize the swag with the name of the project and a cool slogan. Most project owners do not do this. If you do it will carry a lot of social capital when you need their hard work on future projects.

If you have never participated in or led a project before you may want to do some Googling on project planning or pick up a book at the library or online that focuses on project planning 101. It does not have to be complicated. It is worth the effort to be disciplined and develop a plan.

6 important bonus tips for a successful project:
- Be organized, focused and respectful of start and end dates. Understand your resources. Set a realistic completion of goals and reward yourself along the way.

- BE CONSISTENT. If you complete a few tasks everyday it really adds up over a month or year.
- If you hit a wall find a way around it, over it or under it – never let it stop you!
- Be creative and think outside the box.
- Never give up AND Never say 'I can't'! I live every day of my life by those credos.
- Each project is a journey, enjoy it.

How to Generate an Endless Supply of Digital Ideas, Solutions and Prizes
My Favorite Tips and Techniques

'One great idea can change the lives of millions of people.'
Marty Fox

Ideas are priceless. They are your critical first step to building transformation value. The great inventors; Leonardo da Vinci, Thomas Edison, Albert Einstein, Nikola Tesla and Dean Kamen all had to begin with an idea. Digital success is limitless; however, you need an idea to get started.

I am an idea machine and the subject of ideas is interwoven throughout the book. In this chapter I will share some of my favorite tips to help you generate an endless stream of new ideas. Fortunately, by using these techniques and resources I find it very easy to come up with new ideas. My biggest problem is not having the time to act on all my ideas. I am not complaining, as I would rather have to prioritize my ideas than have 'idea block' and no ideas.

Become an idea machine and the world is yours. You can attain business and career success. You will never get bored. You will always have new challenges awaiting.

Let us begin with a few idea tips that work:
Idea Tip - Often the best ideas are the simplest ones. Your idea can even be an idea on making an existing product or service simpler or smarter! Steve Jobs was the master of creating simple and elegantly designed products. It is likely that you are carrying one of his masterpieces in your pocket or pocketbook.

Idea Tip - Look at the world as it can be, not as it is. I call it 'seeing the world on a slant'. Be unique. There are endless opportunities to improve the world using technology. Almost anything can be made better, faster and cheaper. This mindset generates ideas.

Idea Tip - Free your mind from constrains. Generating ideas requires a relaxed open mind. Programmer friends of mine tell me how they can be stuck on a complex problem much of the day and stress over a solution. Amazingly, they go to sleep and then wake up, voila their subconscious solved the problem as they slept. It has happened to me many times as well.

Idea Tip - Think about what you need and want. Your pain might be shared by millions of other people. Build that solution! I describe how the devRant guys did that later in the book.

Idea Tip - When you use a product or service think about yow it can be made better and smarter with technology. The previous sections on mindset and technology described fantasy-like products that became reality. If you can imagine it, you can build it.

Idea Tip - I generate a lot of ideas while listening to music. Music inspires me and sometimes helps pull creative ideas out. Listening to music also makes my writing bolder. Give it a try.

Idea Tip - Exercise in any form always helps me generate ideas and solve problems. Steve Jobs loved to take meetings while walking. I can go on a long walk or run and the ideas just flow like the Nile. Exercise can also improve health.

Idea Tip - Visit cutting-edge technology sites such as Engadget, TechCrunch, Mashable and/or read my digital success newsletter.

Idea Tip - Read annual reports and public filings. You can discover hidden trends and opportunities.

Idea Tip - Visit Digitalgov.gov, NASA and other government sites. They are fantastic resources. You might be surprised how much cutting edge information you can learn on every type of emerging technology. There are also in-depth transcripts from Congressional hearings that can tip you off on trends and new technologies. I used some as research material for this book.

Idea Tip - Watch tech focused television shows. Some have placed their content online, but to get the most value you want to get access to the information as soon as possible and that is often when they premier on TV.

I love every minute of each of these shows and never miss an episode. I watch them with a pad or my mobile device nearby to make notes of interesting tech ideas and companies for further research:

- Xploration Earth 2050 - Chuck Pell is an inventor, artist and futurist. He shows us what's here now and what's coming next. Fox Network
- Studio 1.0 - Emily Chang, the Queen of Tech TV. She does in-depth interviews with technology business leaders - Bloomberg TV
- Bloomberg Best of Technology, Emily Chang does it again! This show is weekly review of the latest happenings in the world of technology and innovations. She conducts interviews with tech movers-and-shakers - Bloomberg TV
- SciTech - Hari Sreenivasan hosts three segments per episode that explore a wide variety of tech and science breakthroughs - PBS
- Click - Spencer Kelly hosts a fast-moving UK originated show that explores the latest happenings in tech, with features on drones, robots, AI, AR, VR and many more goodies.
- Hello World - Ashlee Vance, a talented technology writer, travels world-wide to interview technologists. I wish there were more episodes.
- Henry Ford's Innovation Nation - Mo Rocca and his team share new inventions and interview the inventors, in an entertaining manner. They also have fun multiple choice quizzes and word learning in 'The Mo You Know' segments. The show is great for children and adults of any age - CBS
- CSPAN Book TV - I monitor it for authors of digital technology or business books. They have interesting interviews and the Q&A sessions following the interviews can be informative. You can then purchase their book. Hopefully, someday you will watch me on Book TV discussing Digital Transformation Success Secrets.

Idea Tip - Watch online videos and take online courses. You can experience technology products in action and learn how they work.

Idea Tip - Visit **SlideShare.com** for presentations on new technology. Many are done in an informative and succinct format.

Idea Tip - Watch **Ted Talks** (ted.com) for insights from some of the best and brightest people. Ted is about people and ideas. They can be informative, motivating and sometimes funny. Some can be as brief as 5 minutes and can change your life.

Idea Tip - Attend conferences, meetup's and seminars. They are great for generating ideas and networking. I recently attended **techDayNYC** (https://techdayhq.com) where I met 500 startup leaders and learned about their products and vision. Get out there and network!

Idea Tip - Share a meal and brainstorm with teammates and friends. Many great ideas and startups are hatched while having a meal with people you enjoy being with and letting the ideas flow. Everyone brings a different perspective and each person's thoughts should be respected and encouraged.

Idea Tip - Try mind mapping. A mind map is a way to free-flow ideas from you brain without constraints. I learned about mind mapping when I assumed the independent role as the sole North America business development executive and strategic advisor for a two-person company in Europe. We helped turn Mindman (the name at the time) into a software leader. Mindjet.com has a lot of useful information on mind mapping.

Mind maps can be used to produce ideas, increase productivity and to simplify the planning of any type of an event from a wedding to a complex network. Each mind map has a central thought and free flowing branches that extend from the center. It is amazing what comes out of a mind mapping session. Some mind maps are beautiful art. Google 'mind maps' and you will immediately understand. Mind maps can be created with software or with a box of crayons and paper.

Idea Tip - Use email or build / buy an internal idea crowdsourcing digital platform for your company. Get all your employees involved in idea creation. You can keep it simple by email and offer prizes
for ideas that go to production or you can use a more elaborate system that enables voting on ideas that have been submitted in different categories.

Idea Tip - Read, read and read some more! I am a voracious reader. I always have 3 or 4 books lined up and devour two or three a week. I learn best from reading, rather than attending classes. There is a wealth of content available to you both online and off. I am a big fan of eBooks and reading on my Kindle. I came up with an amazing way to easily convert documents to real eBooks and migrate them to a Kindle for a better reading experience. I featured the story in my 2017 book, 'The Ultimate Business Continuity Guide'. I did not have room in this book for the story but I will gladly email you a copy if you contact me.

Idea Tip - Google Alerts (https://www.google.com/alerts) enables you to discover valuable ideas, contacts and trends. You can set up multiple alerts on different keywords and text strings. It searches millions of online resources on an ongoing basis and pushes precise information, that meets your criteria, to you in real-time or batched (sent in groups during a specific timeframe).

Idea Tip - Hootsuite (hootsuite.com) enables you to monitor social platforms including LinkedIn, Facebook, Twitter, Instagram and Pinterest. It can be used in creative ways such as to generate ideas, monitor competitors, detect potential disruptors, collect information, automatically post content, analyze what people are saying about your company and anything else you can imagine.

Idea Tip - Google Trends is a powerful free idea generating tool that enables you to discover what people are searching for, what is hot and what is not. It has valuable trending charts that enable you to anticipate what and when products will trend in the future. There are many 'peek into the future' uses for this tool.

Idea Tip - Always document your ideas as soon as possible. Ideas are too valuable to lose. The right one can change your life and that of millions of people. When I get an idea in the middle of the night (which I often do) I jot it down on my mobile device or on a paper note pad. Otherwise, ideas just flitter away.

Challenges & Prizes
Challenges and Prizes are fantastic ways to find an idea, test an idea, solve technology problems and possibly win prizes! You can be a participant in existing challenges to help other people solve their problems or host your own to develop new ideas and solutions. In this section, we will learn how, why and where to find them.

A Brief History of Technology Innovation Challenges and Prizes:
Technology Innovation challenges find their ancestry in contests dating back to 1567. Through the centuries, they have helped solve very big problems.

In 1714, The Longitude Act was created by the British Government. Their goal was to find a method to determine the exact location of a ship's longitude. It paid out a series of rewards through the years to many inventors. I like how they creatively tied the amount of the rewards to the accuracy of the solutions. The lowest amount paid was 10,000 pounds, which equates to $1.5 million dollars today. Higher amounts were paid for more accurate solutions. At least one individual, John Harrison, received multiple monetary prizes and the most money of anyone awarded a prize.

Later, the Orteig Prize offered a reward to the first aviator to fly non-stop from New York City to Paris or Paris to New York City. There were several unsuccessful attempts until Charles Lindbergh won the prize in 1927 with a New York to Paris flight.

In our digital age, challenges and prizes have multiplied and are easier to find and participate in. There are many private and government challenges and platforms. These present big opportunities for companies and individuals.

In my opinion, the most famous is the X PRIZE. Peter Diamandis, the founder of the X PRIZE foundation, is also an engineer, physician and entrepreneur. The goal of the X PRIZE is to bring about "radical breakthroughs for the benefit of humanity" through incentivized competition. The X PRIZE competitions motivate individuals, companies and organizations across all disciplines to develop innovative ideas and technologies that help solve big challenges.

In 2004 the Ansari X PRIZE relating to spacecraft development was created. The goal of this prize was to inspire research and development into technology for space exploration. The $10,000,000 prize was won by designer Burt Rutan and Paul Allen for their SpaceShipOne.

Netflix awarded a $1,000,000 prize in September 2009 to team "BellKor's Pragmatic Chaos" that produced an algorithm to improve the accuracy of predictions about how much someone is going to enjoy a movie based on their movie preferences.

Below is a taste of other online challenges as of August 2018. Maybe you will participate on one or more challenges or sponsor your own:

InnoCentive (innocentive.com) is an open innovation and crowdsourcing company that enables organizations (referred to as 'Seekers') to tackle their important business, scientific or technical problems by crowdsourcing solutions from outside or inside their organization. They offer their Challenge Driven InnovationTM methodology, purpose-built platform and global network of problem solvers (referred to as 'Solvers'), as well as accompanying consulting, training and program management services. They are headquartered in Waltham, MA (USA) and London (UK).

Kaggle (kaggle.com) is a community for data scientists and machine learning engineers. It is a great place to learn and share ideas. They also have interesting competitions for talented people that love data and can solve specific problems that are very important to organizations hosting each competition. When I visited Kaggle.com, there were many competitions in flight with prizes ranging from $2,000 to $1,000,000 plus.

If you are an organization, you may want to host a competition to solve a data challenge and if you are a talented data scientist or machine learning engineer you may want learn, socialize and enter competitions.

General Mills Innovation Challenges are created to find best-in-class practices mainly in the fields of technology and food. They believe by collaborating with some of the best start-ups, entrepreneurs, companies and technologies in the world, they can find the breakthrough innovations that will improve how they do business and better meet their consumer's evolving needs.

Challenge.gov (challenge.gov) is a listing of challenge and prize competitions run by more than 100 agencies throughout the federal government. These problem-solving events include idea, creative, technical and scientific competitions in which U.S. federal agencies seek the public's help to solve perplexing mission-centric problems. More than 740 challenges have been run in the federal government since Challenge.gov since it launched in 2010.

Hackathons

Hackathons are the cousins of challenges and prizes. They are a big part of the lore of personal computing from the dawn of the computer revolution until today. Some of the most famous programmers in the world participated in hackathons. The original concepts for successful products such as Twitter (internal company hackathon), GroupMe and EasyTaxi were born during hackathons. Many other products were developed, became startups and ultimately were acquired by big companies.

Hackathons are often focused on developing software solutions, but not always. Recent hackathons include, robotics, IoT, 3D printing and design.

Hackathons are an effective way to solve problems, learn, socialize, collaborate, work in a team environment and possibly win prizes. They can go from a few hours, 24 hours, 72 hours or longer, so stamina and lack of sleep are to be expected. Food and fun are usually provided by the sponsor(s). A big benefit to participants is when industry experts are available during the events to provide guidance and suggestions. The devRant.com founders, featured later in the book, volunteered as experts in a recent well-attended hackathon in New York.

Hackathons are not limited to startups. Anyone can participate or sponsor a hackathon. More and more enterprise companies and VC's are sponsoring events. They benefit with goodwill and perhaps access to a new technology product that can rock the world.

DevPost (https://devpost.com/hackathons) has an impressive list of upcoming hackathons which includes the amount of prize money being offered by the organizers.

If you are considering creating your own hackathon, I suggest you read **Joshua Tauberer's** online post **'How to Run a Successful Hackathon'** - (https://hackathon.guide/).

Why Digital Strategies are Different and How to Profit from Them

In this chapter, we will discuss four of my favorite digital strategies to transform products and services. Each strategy is malleable and can be used alone or in combination with other strategies to provide extreme customer happiness.

As you read through this chapter think about how you can apply some or all the strategies to your existing and new products. They might even help you create 'competition-less' markets all to yourself.

Transform physical products to digital - Transforming a physical product made of atoms to one comprised of bits and bytes can provide many advantages such as:
- Cost - When products are digitized to bits and bytes the cost of reproducing them is as close to zero as you can get.
- Functionality - digital products, especially when they reside in the cloud, can easily be updated and become more useful to all customers instantly.

Some products become better by being fully digitized, such as maps, books and magazines. Other products such as sports equipment and mobile phones become better by remaining in physical form and adding intelligence software, sensors and connectivity. We discuss many such products throughout the book.

Bundle - Bundling individual products together can provide extra value for the customer in functionality and price savings. It can also simplify the choice between buying separate products when they can have multiple or all of them for a reasonable price. It can be a differentiator for you. If a competitor is selling a widget for $100 and you are selling the same widget in a bundle with two other valuable widgets for the same price all things being equal, it would make sense for the prospect to choose your product.

Because of the unique advantages of selling bits and bytes, software can be bundled with hardware or other physical products and distributed at no additional cost to the consumer.

For example, my son learned programming at a very young age in part by writing computer games. His games were widely distributed online and in one case a popular European magazine gathered a variety of games and distributed them on a CD attached to the cover of their print magazine. It was a win-win scenario. The magazine had a higher perceived value for buyers and my son and the other game programmers included on the CD received huge distribution of their games, which led to additional opportunities.

I used a similar strategy to increase the value of many of the books I wrote for programmers. I included a disk containing the programs described in the book. Programmers could use the routines without having to input them. This saved them significant time and energy.

Unbundle - If you are selling a large comprehensive solution it could be beneficial to break-out / spin-off a feature that your data analytics shows is being used by a high percentage of users. The most popular example would be Apple unbundling music which led to the wildly popular Apple Store.

Steve Jobs had the idea of unbundling singles that were previously bundled on albums. Previously consumers were forced to buy all the songs even though they may have only been interested in one or two. By digitizing and unbundling the individual songs, they could be sold for 99 cents each rather than $8-$12. Consumers loved the idea and the rest was history.

I have successfully unbundled digital content and precisely targeted it to niche audiences. Readers love it as they get the solution with the least amount of noise.

Customize - Customizing products or services adds value for the user and additional revenue for the creator. If you sell a commodity product you may be able to customize it and make it a premium product. Digital tools and technologies often make customizing simple and practical, whereas it would be impossible or prohibitively expensive in the industrial world.

Many of the digital technologies and tools we discuss in the book, can be drivers for customizing your products or services. For 'soft' information customization, meaning providing users with the precise insights they need, artificial intelligence, IoT devices, API's, databases, communication tools and alert products such as Google Alerts and Hootsuite can analyze torrents of data and automatically match and communicate it to the needs of individual users in real time or batches.

3D printing is empowering the creation and distribution of customized 'hard' physical products from prosthetics to jewelry to golf clubs and most any other product. Software drives 3D printers as you learned earlier in the book. You won't even need to own a 3D printer to build a customization business, you could simply use a shared one in the cloud.

Customizing t-shirts and other clothing is as simple as uploading design criteria to a platform provider. You can even do cost effective customization on an individual basis.

We will discuss customizing and repurposing content in the upcoming chapter, 'How to Leverage Content, the Currency of the Digital World'.

Find Opportunities Beyond Your Industry- Widen your scope. Use automated digital listening tools described in the earlier chapter, 'How to Disrupt the Distruptors'. Read a lot to monitor distant niches for opportunities to disrupt them and to discover how they might disrupt your niche.

Also, keep a keen eye out for what other industries are not doing. This will help you find companies in pain and you can alleviate it with your solutions. Find the prospects, show them what you have, improve with free feedback and go disrupt a niche or two.

Find Opportunities Internally - Often companies use a strategy of digitally transforming marketing, sales, customer service, and some other operational units. However, they stay old school with their products and that is a big, big mistake. Core products and services must be enhanced. For example, think about the education industry and how people are now learning on the Internet, for free or using a freemium model. This must be on the radar for expensive colleges and universities. They must become better, faster and cheaper.

There are likely digital low-hanging fruit opportunities in your organization that have been overlooked. How can you automate time consuming and error prone systems that are destroying your second most valuable asset - data? I have identified and improved processes in every company I have worked in.

Carve Out Your Own Unique Market - digital products are inherently flexible and lend themselves to unique new value-laden customer solutions. Rather than trying to incrementally better the competition with new features or race to the bottom in price, try to create entirely new products that separate you from the competition and open new markets. That is where ideas and imagination become critical.

All the software products, books and videos that I have developed have been unique. They did not really have much competition. Some were first movers, such as my Video-Vu videoconferencing tool and CartEasy eCommerce engine. There were other books for programmers and digital technologists so I approached mine from very different angles. I think the book you are reading is a good example of my unique real-world hands-on style.

Case Study: How Digital Strategy Transformed Getting from Point A to Point B
When traveling to a destination we are not familiar with, we need directions. I am exhibit A when it comes to getting lost even on short trips.
Travel directions have been transformed through the centuries. In the beginning, the stars were used to guide us at sea. We needed a better way for shorter land based trips and physical maps made of paper or plastic were developed. For thousands of years these were the dominant formats for understanding how to get from point A to point B.

Perhaps you remember fumbling with a 3'x3' Rand McNally type map while driving on vacation or on a business trip. I am not great with directions and even if I prepared in advance and highlighted the route somehow, I would miss highway exits and turn down one way streets the wrong way. Maybe it was just me but I was a mess and using those maps while driving was a recipe for stress and being late to appointments.

Due to the efforts of NASA and some cool technology that used satellites that triangulated the point of an object on earth, even if it was moving like a car, the magic of GPS (Global Positioning Satellites) was born. It was amazing for most of us, unless you owned a company that produced paper maps and you did not see GPS disruption coming.

Enterprising entrepreneurs formed GPS companies and began producing plastic units that were far more valuable than paper maps. GPS units typically cost hundreds of dollars but they were well worth the price for a large part of the driver universe - especially direction clueless me. GPS units flew off the shelves!

Unless you were there it's hard to imagine how cool it was to not have to reach for a paper map and have a pleasant voice tell you where and when to turn all the way to your destination. I was once driving with my friend Marc, who is not a technologist, and he made an insightful remark about technology. He said the best tech inventions were the GPS and automatic toll scanners like EZPass. I would agree. Each made the customer's life better and faster.

At the beginning, I did not trust my GPS. I recall a vacation drive with my family from New York to Maryland where I doubted the GPS and my young son got mad and threw it on the floor to protest my doubts. By the way, the GPS was right. I became a believer and that was the last time I doubted it.

In effect GPS companies disrupted paper map companies that did not have the foresight to morph into GPS. Many of the paper map companies that were industry leaders are no longer in business.

Hold on though. The story of directions was not quite over. Technology always marches on and as I have stressed throughout the book we must always have our radar up.

The plastic GPS units were great but there were some drawbacks. They were expensive so some people could not justify the expense if they did most of their driving locally, which is the case for most drivers. One or two long distance vacations a year might not be enough incentive to shell out a couple of hundred dollars. Also, roads changed and the GPS companies would charge for the updates.

At the same time as GPS was the rage, the mobile phone industry was exploding in popularity.

Telecommunications and the mobile phone were not even what I consider a tangent industry to maps and directions, until they were. Remember, distant industries can provide opportunities for you if you keep your radar up.

Millions of us were carrying mobile phones as they became smaller and cheaper. Mobile carriers included basic software services and email functionality which whet our appetites for more. With the convergence of mobile and improved computer processing power, mobile devices morphed into tiny supercomputers. That opened opportunities to add additional software functionality and access to web sites.

At first many of us would visit MapQuest.com and although the experience was not perfect, we could access directions on our phones for the first time. We already owned the physical mobile device so it should have been a warning for companies that produced the dedicated physical GPS units. Importantly, because the mobile software was centralized in the cloud it could easily be updated with the latest road changes without distributing it to each endpoint device. It was the beginning of the end for companies that formerly disrupted the paper map companies.

The iPhone was becoming wildly popular for its design and status. It was a big reason for the disruption of Blackberry's which had been the coolest status symbol at the time, but you better not let your kids see you carrying one in 2018.

Steve Jobs, CEO of Apple, and one of our great innovators started having his team add apps to the iPhone. Apps, as you know, are tiny software programs that run on mobile devices. Travel directions were offered as an app, but Apple did not do directions well.

Jobs struggled with the decision whether to allow developers to offer apps that ran on Apple devices. At first, he was against it for security reasons but finally decided to give it a try. The rest is history as there are millions of apps available in their ecosystem through their Play Store.

Google did the same with Android and opened it up from the beginning to software developers. A tiny company in Israel developed a directions app that became very popular. It brilliantly took advantage of the crowd which is also the secret of many companies we speak about in the book, such as Wikipedia and LinkedIn. The tiny company was Waze and it offered its app for free.

As you will learn in the next chapter, free is tough to compete with. It was data gathered from users that was the special sauce for Waze. Using gamification, drivers were rewarded with points for reporting potholes, red-light cameras, cars stuck in the shoulder and hundreds of other events.

The information Waze collected was instantly made available to other drivers. The data was beyond gold. I count myself as one of the drivers that ditched my plastic GPS and only use Waze. It worked out pretty well for Waze as well. They were acquired by Google for one billion dollars. Google went on to add many other digital features to Waze that took it to a new level. They also added intelligent ads based on location data which was another win for customers, vendors and Google.

Digital transformation strategies will continue to proliferate. There are billions of people online and you have so much more flexibility when selling bits and bytes as opposed to atoms. You might even discover an entirely new way to profit from the unique characteristics digital has to offer.

Why Digital Business Models are Unique and How to Profit from Them

In this chapter, we will discuss my favorite digital business models. Some of these models are unique to digital products and others can cleverly leverage digital and physical products. I will intersperse some of my personal successes using these models.

As you read this chapter start thinking how you can apply these to your current products and services or new products you can invent that can benefit from these models. Possibly by using one of the digital business models below, a combination of models or by inventing your own digital business models, you can add new revenue streams.

I encourage you to experiment, think out of the box and test. Pick and choose, mix and match. Digital business models and digital products provide the opportunity to experiment with little downside risk and huge upside possibilities when you discover the right formula.

'Free' Digital Business Model
Question - How much do you think each of these products or services costs the end user?
Google Search?
Google Docs?
Google Sheets?
Twitter?
Facebook?
99% of the web sites you visit?
Physical snail-mailed trade magazines?

The answer is zero. They are free! Welcome to the free lunch. Sure, we can nitpick and say you pay micro-cents for electricity and bandwidth, but essentially the products I listed are as close to free as you can get.

So, the obvious question is, why would a smart company like Google give something away for free? What is in it for Google or you for that matter? My best answer is Google makes a nice profit giving you products. The good news is you may be able to use free in your digital business.

Free has a long history as a powerful physical and digital business model for products and services. Giving products away for free is a viable way to make money and build credibility that can lead to lucrative opportunities. For example, Gillette on occasion gave away razors to create a marketplace for their blades.

I have successfully used free and its cousin freemium, which we will discuss later in this chapter, many times to sell products and services. On one occasion, I foresaw that free would be popular on the Internet, so in 1995 I registered the FreeProducts.com domain name. I used it to give away my content in a two-step process which led to book and product sales. Later I sold it to a company that had the idea of building a business based on the free business model.

Here is why free works:
- Free is a magical word to buyers. When people are offered something for free they generally take it if they think it is a legitimate offer and the product has value.
- It is tough for competitors to compete with free. Even if a competitor charges 1 cent for a product or service there will be a sharp drop-off in participants between even that tiny cost and offering a product for free.
- Free can be the first-step to downstream monetization integrating models such as freemium and subscriptions, which we will discuss in a minute. Once you get people hooked on the value you provide, you build trust and can sell them on the next level of product or service.
- Free has launched some of the most successful companies in the world.

A few examples of companies that have built their business using free are listed below:
Google - Think about it, how often do you search Google or use their applications, which are great by the way. When was the last time you cut Google a check for the privilege of using their search engine?

Google generates most of their revenue from the little text ads that you and I and most of the planet click on billions of times a day while searching for everything under the sun. Each time we click on a link it is 'ka-ching' for Google. Those clicks add up to billions of dollars of revenue. The little text ads on your desktop and mobile devices enable Google to work on other moonshot business experiments such as autonomous vehicles and automated AI assistants.

The way it works is sellers bid for your attention using the Google Adwords self-service ad buying platform. Google offers sellers the targeted buyers that are interested (demonstrate intent) in a product or service by clicking on the little text ads in the Google search engine or on a link with a third-party affiliate site.

Depending on the product or service, clicks prices can range from 25 cents to $50 or more. Imagine that, more than $50 for a single click. If the person who clicks becomes a buyer of an expensive product, it makes sense. For example, a lawyer that can sign a client for $50 or $100 would be ecstatic. The cost of the click is miniscule compared to the profit. If you are selling magazine subscriptions 25 cents a click would make more sense.

Microsoft - The Bing search engine works in a similar manner to Google. Microsoft used to make a fortune selling software on disks and CD's back-in-the-day, but those days are long gone. Microsoft is all-in on the new digital way of doing things for a reason - survivability. Their flagship product Office365 lives in the cloud and is sold on a subscription basis. They are also a serious player in artificial intelligence, virtual reality, augmented reality, quantum computing and cloud hosting services.

Red-Hat - They were a pioneer in giving away their version of the popular open-source Linux operating system, which is favored by highly technical organizations. Their business model, as often is the case with open-source software, is to provide service and support. Businesses that can save time installing and tuning Linux are happy to pay Red-Hat. They have grown a healthy business predicated on monetizing free.

Facebook - Social networks for the general public rarely charge users fees. They typically make money from advertisers that want your attention. The targeted ads that show up in your news-feed generate revenue hand-over-fist.

Specialized social networks, such as **devRant.com**, featured later in the book, successfully use the free and freemium membership subscription model.

I have been giving away free valuable content in the form of white-papers, webinars, podcasts, videos and speaking engagements for many years. I was even doing it pre-Internet. This 'two-step-process' (free to revenue) earns me attention and credibility. That leads to opportunities to help companies through my brand and to sell paid content in many forms. I speak extensively about creating and valuing content in the upcoming chapter 'How to Leverage Content - The Currency of the Digital World'.

'Freemium' Digital Business Model
Freemium can be a very good strategy to use with digital offerings such as software, media, games, online-communities or web services. I have enjoyed freemium success on many occasions, as I will describe below.
Freemium works well as a 2-step process. You provide a basic product or service for free with the goal of deriving value, often monetary, at some point in the customer journey.

Freemium can take on different forms such as:
1 - Freemium - Full Featured - Limited Time or Use Trial:
This model allows a user to try a product for a limited time-frame or number-of-uses. For example, allowing users to try a product for 30 or 60 days is popular for business software. Games can be used for number of plays, perhaps 200 or 300.
The goal, whether timed or number of trials, is to give the user enough time so they get hooked on the value of the product, but not too long. At the end of the trial the user must purchase a license key and register through an in-app purchase or other means to continue using the product or removing the watermark. I have purchased many products because I enjoyed a trial version tied to the freemium business model. For example, Scrivener, the software I used to write this book allowed me to try a full version for free for 30 days. I enjoyed the value so much that I purchased a license after only 3 days of using it.

The more you get your freemium products into the hands of users, the more sales you can make. Often it is a numbers game. Fortunately, digital products consisting of bits and bytes rather than atoms allow you to cast many seeds into the wind with minuscule cost, if any. On the other hand, imagine the cost of mailing thousands or hundreds-of-thousands of physical products to prospects world-wide. You would need a high sales rate to make it worthwhile. With digital, even a small percentage of sales to a huge audience can be lucrative.

A slant can be allowing the user to try a full featured version without a time or use limit but a water mark is placed on images or documents created with the free version. If the users like the product they will want to remove the watermark before distributing their content, as it makes it much more professional looking. If they distribute it with the watermark, it can be a great advertisement for the company providing the tool.

Canva.com, which I use extensively when designing my books, has free and low cost products you can choose from. You can upgrade if you want more features than these that are offered in their free version or if you would like more elaborate images.

I have used freemium extensively over my digital business career. A memorable example was being asked to be an adviser and sole North American representative for a tiny European two-person startup software company. Using freemium and guerrilla marketing I could widely distribute freemium copies. MindJet became wildly popular with students, colleges, businesses and the government. We marketed it by offering a 30-day full trial. Like clockwork around the 28th day people would pay for a license or multiple licenses. Many would put a special note with their order imploring us that they desperately needed the key to continue using the product without interruption. The product offered great value and the new buyers would spread the news to their connections, which led to even more sales. We never needed to pay for advertising.

Another slant on freemium is to limit the number of records entered in a database or app. Many digital companies ring the register by limiting capacity. For example, Dropbox became a billion-dollar company offering free accounts that turned into paid accounts at a certain threshold.

When my freemium Apple iCloud account recently surpassed 4.6 GB of my priceless pictures, I received notifications that for 99 cents a month I could upgrade to as much as 50 GB. For $2.99 per month I could get even more space. As you know from reading this book, digital storage space and bandwidth get cheaper every day so companies are making a huge profit on selling storage and buyers have a safe and secure place to store all of the pictures and documents that they create. It is a win-win scenario!

In-app purchases of virtual products can be very lucrative for online games. Virtual products can provide super powers that can save gamers time and help them rise on leader-boards. For example, Farmville built a huge business selling virtual tractors and other farming tools for their game. In fact, they were selling more virtual tractors than all the physical tractors sold world-wide. The virtual product craze that began in Japan is now a global phenomenon.

Gamification works well to engage users especially with games, social sites and surveys. Participants can be awarded points, which can be redeemed for gift cards, virtual products or swag. Tying points to leader-boards can also be effective. People enjoy competing and the social recognition of being at the top of a leader-board is cool and to many people is worth more than money.

Points can be advantageous for both the company awarding them and the receiver. It is amazing what people will do for points or t-shirts. Fred Lebow, who launched The New York Road Runners Club and the NYC Marathon, wrote a great book named 'Anything for a T-Shirt'. As a racewalker and runner, I admit I would do just about anything for a t-shirt.
I know of companies that award 1,000 points for doing a survey that can take 20 minutes, even though the cost to the company awarding the points is 25 cents and the recipient might need to earn 25,000 points to get a $25 gift card. If

someone stopped you on the street and offered you 25 cents to do 20 minutes of work, would you do it? Online points are big business and one to consider.

Fitbit is one of my favorite gamified platforms. I have a friendly daily step competition with my wife and friends. The platform is at least partially responsible for the 23,341,748 steps I have as of August 2018!

2 - Freemium - Subscriptions and Limited Features:

The subscription model is growing rapidly. Hundreds of millions of people participate in online subscriptions. Amazon Prime, Netflix, Salesforce, Microsoft and thousands more companies are now using subscriptions to drive revenue.

Advantages of this model include ease-of-use for the subscriber and recurring income for the business. You can offer your basic service for free and enhanced benefits for subscribers to incentivize them. I know of niche social media sites where thousands of users happily pay $2.95 - $7.95 per month to participate in the benefits and in the continuing success of the site. Some people want to give back to online communities that improve their lives. Social media sites, especially niche ones, can provide great value and often become a vital part of their everyday life.

Where paying a fee of $100+ for a one-time purchase may turn a lot of people off, paying 99 cents to $10 or more a month is much more palatable and less of a buying decision. The subscription model is also enticing to users when they can unsubscribe at any time without penalty. Providing that incentive can remove friction from the buying decision.

If you are providing value that you promised people, they will keep the subscription going month after month. If it is a small monthly sum, unless they are very disappointed or find a competitor, they will not cancel their subscription. Some people feel it is not even worth the effort to unsubscribe.

It is important to provide users with multiple easy ways to begin their subscriptions, such as through the use of PayPal, Venmo or other electronic services. If you have an app for your service, in-app subscriptions can be set to automatically bill users on a regular basis, such as weekly or monthly. Any way you can make it easier for them counts toward additional sales.

Subscriptions can also be part of your exit strategy if you decide to sell your digital business. Companies and venture capitalists acquiring or investing in a business like to see healthy recurring income to properly value it.

Affiliate programs enable you to drive sales of your products by creating relationships with web sites, blogs or independent sales reps. Affiliates provide digital real estate on their virtual real estate and when visitors or prospects click through and purchase your product or create other value, the affiliate receives a commission.

Ways to work with online affiliates can include:
- Cost per click (CPC) - commission is generated each time a link is clicked. Google AdWords and AdSense use this model extensively. If this interests you, visit Google for information.
- Cost per action (CPA) - commission is generated each time a certain action occurs, such as a sale or a lead being generated by the visitor filling out a form or calling a toll-free phone number.
- Cost per thousand (CPM) - commission is earned for every thousand page views. This model is generally used with display ads. If you have a very high trafficked site this can be profitable.

There are many affiliate networks such as **RevenueWire** (revenuewire.com) and **Commission Junction** (CJ.com) that take care of the back-end infrastructure. These networks offer hundreds of affiliate opportunities. Most offer CPA commissions tied to sales or leads.

You can incorporate the above techniques through the aforementioned networks or implement them directly using third party software. It would then be your responsibility to contact affiliates directly. It is more work to sell it yourself but you can potentially make more money by not paying a 'middleman' network. I have done it both ways with success.

For example, I have used home grown CPC in directory type sites I create. It really is not complicated. I drive traffic to these sites and then approach companies that would benefit from representation and access to visitors. It is a two-sided platform with no physical product involved. Often potential advertisers contact me!

I always make it clear to visitors where there is a paid advertiser and I only contact advertisers that could benefit my audience. You must maintain credibility. There are now laws where you must let visitors know that you receive commission, so it is important to abide by those.

Channel Partnerships are a step beyond affiliate programs and for the right companies they can provide you with a good revenue stream. Channel Partnerships are one of my favorite types of digital partnerships and I have successfully participated in them on many occasions. The results have varied from little to great revenue. It depends on the company and your ability to generate prospect interest. Remember, not every relationship has to be a home run.

For example, I have a wide network of online relationships with many C level executives and readers of my books, white papers and other content which allows me to help a variety of companies. I have the ability to assist technology conference organizers and companies trying to fill advanced technology positions with hard to locate candidates. If you can make connections, you can make serious money.

Participation in channel partnerships can be as little as simple referrals that can generate 5% commission to lead generation and setting conference calls with a company rep that can generate 20% or greater commission. On a high-ticket item, this can amount to significant revenue for little work.

The best partnerships are ones in which you have knowledge of the niche and can reach the type of prospect that matches the company's customer profile.

There should be no cost to you to participate in channel partnerships. In some cases, the channel partner will set you up with a company email address and a phone extension for branding and credibility.

Sponsorships can be lucrative if your digital product can provide sponsors access to highly desirable prospects in their industry. I am very particular in choosing sponsors. I only consider using sponsors I respect. I choose companies that can add value to my readers and visitors with their products or services. Sometimes they also provide special discounts and swag to my contacts.

Companies often sponsor games and give them away for free to their prospects and clients. Alternatively, some popular third-party games include relevant paid sponsorships. These product placement sponsorships can be changed dynamically so if one sponsor drops off others can easily be added. Product placement has been lucrative in the movies and television for decades and now is prevalent online.

Sponsorship sales to companies can be an easier path to monetization than game sales to individuals.

Elastic 'demand' pricing' is used by many companies, including **Uber**, **StubHub** and **Kohls**. When demand goes up, pricing can rise accordingly and the opposite occurs when demand is weak.

In the physical world street vendors selling umbrellas or water do this all the time. On a rainy-day, umbrellas are coveted and demand a higher price. Ice water on a hot day is another example of a product where value increases due to the high temperature.

In the digital world, it is easy to automatically adjust prices to demand in milliseconds. Kohl's has incorporated big data, AI, chatbots and predictive analytics to create customer delight and it has paid off for them. You may have noticed the LED pricing screens throughout their stores that can be dynamically changed depending on demand all from a central location.

I was fascinated by the screens every time I was in Kohl's so I did some research. I found the supplier of the wireless digital signs that Kohl's uses is an innovative company called Altierre Corp. (http://www.altierre.com), located in San Jose, California.

Altierre describes how excess use of paper price tags and signage, ink cartridges and other materials add up to a $50 billion annual problem for some industries in the United States alone. Imagine the manual labor, mistakes and resource waste that goes into changing thousands of products at a small to mid-size store without automation.

To reduce that waste, save trees and open real-time data-driven opportunities for clients, Altierre offers digital tags and other digital IoT devices and sensors. It is a win-win scenario and the sustainability benefits are lovely icing on the cake!

Custom Pricing: Whereas elastic pricing rises and falls depending on outside forces, custom pricing can be used depending on what is known about the customer. In the physical world, certain assumptions are made by stores aligning to broad zip codes or other somewhat effective data points.

The digital world takes it to higher level. Companies know much information about us and many adjust prices depending on who is looking at a page or a product. Reverse IP'ing, geo-location, big data and AI can provide valuable high precision information to match the product to the visitor. Not every visitor to a web site is seeing the same price. Algorithms are dynamically adjusting the pricing to make it more enticing to each visitor.

Platforms are core to the Internet and can be very beneficial for the platform facilitator and participants. They can bring together parties that have friction finding each other in the physical world. Another way to describe platforms are digital marketplaces.

If you have an idea for an under-served market or are in a niche that has friction and does not have a platform, building a peer-to-peer platform might be an interesting opportunity to consider.

Platforms benefit from the network effect. The more buyers and sellers that participate, the more desirable a platform can become. The digital advantage removes most of the costs and geographic limitations of physical marketplaces. The types of platforms that can be built are almost unlimited. Real estate, dating, ride sharing, retail, etc.

The company that builds a platform can monetize it in many ways. For example, there can be commission on each transaction, membership fees, third-party product sales, branded products from the facilitating company, fees to advertise products or services, etc.

You do not have to be a technology company to build a platform. There are off-the-shelf peer-to-peer platform software solutions, such as extensions to Magento (magento.com) as well as Near Me (near-me.com), Sharetribe (sharetribe.com) and Marketplacer (marketplacer.com) that can make development and implementation much easier than building from the ground up. Each of the companies describe case studies and features on their web sites.

Two popular types of platforms are:

    1) One-Sided (Single-Sided) platforms such as customers interested in buying sneakers, cars or jewelry.

    2) Two-sided (Multi-sided) platforms bring together two parties. They can match:
- People with other people - such as dating or business
- People with things - such as jobs or real estate
- Machines with other machines - connecting IoT type devices using middleware

Here are a few examples of how two-sided platforms are changing the world:

Craig Newmark, a programmer with an idea and some available coding time, started **Craigslist** (craigslist.com) on a shoestring. It has been instrumental in disrupting newspapers lucrative classified ad revenue streams. Notice how simple and effective it is. Like Google, it is light on graphics and heavy on providing value.

Kickstarter (kickstarter.com) changed the way products get financed, made and brought to market. Many great products we now use would never have seen the light of day without this crowdsourcing platform. It is great when supporters show their enthusiasm for a person's idea and it becomes a customer delighting new product. In addition, the individual or team creating the product limits risk and does not have to give away ownership.

Patreon (patreon.com) is a platform that makes it easy for artists, writers, podcast and video creators to build a sustainable revenue stream and retain 100% ownership in their creations. I am using Patreon for digital clients.

    The breadth of subjects is a wide range from educational and scientific to games, music, photography, comedy and everything in between. For a small 5% commission Patreon assumes the complexity of creators getting paid from fans of their work.

    Patreon, like many successful digital enterprises, was created to fill a personal need, as are many successful digital enterprises. Jack Conte, a popular YouTube musician created videos that millions of people enjoyed but he was making little money from his efforts. In 2013 he teamed with his college roommate, Sam Yam, and built Patreon, a platform that enabled him to generate revenue from his creations. He then shared Patreon with creators worldwide to enable them to do the same.

As of September 2018, there were 100,000+ creators on the Patreon platform. Some are making well over $150,000 annually. They can establish subscription levels that give members various levels of rewards. The 2018 projected platform earnings for creators is $300 million.

LinkedIn (linkedin.com) is the largest platform for professionals. I use it to connect with peers and I find many new business and branding opportunities. The platform enables me to share my content and learn from like-minded professionals. Companies use it extensively to find their next employees. By the way, please connect with me on LinkedIn. Mention you read my book, as I get many connection requests and must be careful to keep my network focused.

Alibaba (alibaba.com) has changed the dynamic in finding product sourcing companies and manufacturers in distant countries. It addresses the needs of startups to enterprise companies and everything in between. It is now branching out into other areas such as facilitating payments between buyers and sellers.

Facebook, Twitter, Instagram, Pinterest, devRant and Reddit connect people by location, schools or professions. The power of these social platforms can change lives, careers and governments.

ZipRecruiter, Indeed, TaskRabbit, RepHunter, Manufacturers Rep, Amazon Mechanical Turk (https://www.mturk.com/) **and Freelancer** (https://www.freelancer.com/) connect people with full time, part time or contract opportunities. Typically, you complete a profile and when an opportunity matches your skills you receive a text or email with a description and the opportunity to apply.

Behance (owned by Adobe Systems) has a user base of more than 10 million creative people. It is a fantastic site to showcase and discover creative digital work. It consists of galleries, tools, jobs and instructional videos.

The App Store, Google Play and SoundCloud have unlimited shelf space and feature millions of apps and songs. Many use the free and freemium digital business models. Increasingly, we are discovering 'there is an app for that'. If not, maybe you should satisfy customer needs by creating the app! Some huge businesses were started with a simple freemium app and in-app purchases.

eBay is the ultimate garage sale. It squashed the friction of finding that needle in a haystack product that can be gold to a buyer and 'found money' to a seller.

Tinder, Match, OK Cupid and eHarmony are dating sites that facilitate the type of a one in a million chance occurrence that took place in the industrial age Tom Hanks, Meg Ryan movie, Sleepless in Seattle (no spoilers here). Simply swipe right or left and you can find that special someone. Dating platforms can greatly reduce the friction of meeting people located anywhere by destructing geography. They significantly increase the pool of potential mates. In the long term the sophisticated matching algorithms they use hopefully will increase the longevity of marriages and provide a lifetime of happiness.

Google (http://google.com) has been extremely successful matching searchers with products, services and content. Advertisers love the 'intent' factor when someone is searching and therefore, Google's little textual ads generate billions of dollars.

SalesForce
(https://www.salesforce.com/products/salesforce-platform/)
is the world's largest customer relationship management (CRM) platform. It has its own cloud infrastructure and programming language, Force. Many enterprise products such as Fusion Risk management are in their ecosystem.

There are too many features to include in this space so if you have interest, check it out.

Predix (htttps://www.ge.com/digital/predix) is a platform from GE that extends industrial applications on a secure and scalable platform. Industries as diverse as automotive, aviation, manufacturing, mining, oil and gas, power utility and renewable energy leverage the platform. Many of the applications cater to Internet of Things. For example, GE has famously used it to analyze torrents of real-time data from its jet engines. That enables predictive maintenance and safer planes.

Lyft, Zipcar, Peerbuy, AirBnB are part of the 'The Sharing Economy', which is changing economies and disrupting industries. They are powered by platforms and for the first time makes access rather than ownership possible in many industries.

Nest (http://nest.com) is creating a wonderful platform that connects IoT devices to build smart homes. They partner with 60+ automobile, home appliance, utilities and digital software companies to provide extraordinary value that no individual company would have the resources to create.

IFTTT and Zapier are great examples of easy-to-use platforms that connect IoT devices and systems. I have used them to connect smart lightbulbs to Google Home which further connect to a mass notification system. There are thousands of uses for these platforms rather than using more complex API's.

Digital ecosystems are typically built on top of platforms and leverage the infrastructure and audience. If you produce a product or service that maps to a popular platform, leveraging an ecosystem can be important to your success.

The apps in The App Store and Google Play are part of an ecosystem, as are songs on iTunes. The 4,000 plus apps that live in the Salesforce store can be of value and sold to the platform participants. Posting products on eBay or selling items on Amazon are ecosystem opportunities.

I have both built a platform and launched digital businesses as part of ecosystems on existing platforms. When I write content or develop programs, I seek platforms where they will provide the most value to the participants. This greatly enhances awareness and the reaching of targeted prospects.

ECommerce Carts and Payments solutions are flexible, customizable and visual to fit your product line. They can even process payments and provide alerts to high-risk buyers. If you are shipping physical products to a global buyer base, they can calculate the shipping rate to various countries and print a Post Office ready shipping label.

Magento, which was acquired by Adobe, (magento.com) and **Shopify**, (shopify.com) are two robust, easy to implement and popular solutions used by millions of stores. They both have good reporting features. Each is easy to implement and if you need assistance there is an ecosystem of developers and consultants that can help.

Getting paid is important. Whereas years ago, it was very difficult to take payment online, now there are many options, including **PayPal, Stripe** (stripe.com)**, Square** (square.com) **and Authorize.net**. Back-in-the-day when I started developing and selling my software online I had to sub-lease a credit card machine from the owner of a shareware company in New York and pay a commission on each sale. It was my only option at the time. Things are far easier now.

The back-end payment processor you are using will seamlessly confirm the payment is authentic and not fraudulent. You will receive a ka-ching email with details of the purchase and the processed amount of money.

By the way, I have used a ka-ching email for many years. When I receive an order, a ka-ching sound plays and I smile. I got the idea from Jeff Bezos. I read somewhere that when he started Amazon he received a ka-ching email every time he received an order. Imagine if he did that now! It would be a steady ka-ching tone 24x7. I receive a lot of ka-chings, but I am confident Jeff got a few more than I get.

How to Create Value Fast Using MVP's and Pivots

A unique characteristic of creating digital products is the ability to develop them in phases without committing to the final product up front. In this chapter, you will learn about Minimum Viable Products(MVP's) which many successful digital companies and inventors use to create products and services people want and need. I have successfully used MVP's and even added my own slant stressing 'value' rather than 'viability'. I call my version Minimum **Valuable** Products. More on my reasons for the slant later in the chapter.

MVP's incorporate enough features to satisfy early customers so they will use the product and provide valuable feedback that can suggest improvements the creator can implement without the expense and risk of building all the bells and whistles up front. There is a great deal of logic to this methodology but there are also some important things you must be aware of.

Developing extensive features increases costs and risk if the product fails. MVP's remove the risk of assumptions and gut-feelings, which are often not accurate. Instead MVP's use hard facts gleaned from customer generated feedback data.

MVP's enable you to prove there is a market before you spend a lot of resources building out the full-blown product and before you spend money marketing a product people do not want. This has worked for me with software I have created for external and internal products as well as for books I have written and published. It saves me a lot of time and money and optimizes revenue.

I have found that MVP's also remove or minimize crushing habits such as 'Paralysis by Analysis' and 'Paralysis by Perfection'. MVP's allow for fast cycles (iterations). They keep the ball rolling. Using a customer focused agile development methodology and rapid-application-development tools you can implement cycles and improvements very quickly. This enables getting better products to market faster which can mean disrupting an industry.

Early adopters are often the first to get their hands on a MVP. These are people that enjoy trying products as soon as they are available. They are not concerned with every bell and whistle. I am such a person. Get a cool 'unfinished' beta product in my hands and I will play with it and I will give you valuable feedback.

For you the value of receiving and listening to user feedback is that you are building what the customer wants, not what you think the customer wants. The chapter, 'How Two Technologists are Connecting Programmer's Worldwide' is a great example of achieving success by listening to what users want.

There are many ways to gather user feedback including interviews, surveys and questionnaires. Watching users in action can provide a wealth of information that will help you carve out your best product or service. If a product is not doing well or if users demonstrate a demand for a new feature and benefit, a pivot (change in direction) may be appropriate. You can revamp your product to meet the needs identified or you may find other sectors that can use your current features.

A classic pivot was that of a struggling social media platform that was having difficulty competing with a slew of similar platforms. They analyzed metrics gathered from their customer feedback loop. The data clearly indicated there was one imaging feature on their platform that users raved about and more important was used far more than any other feature.

The company decided to pivot their social media platform to an image focused media platform. It worked spectacularly well! They quickly became one of the most popular platforms on the planet and were acquired by a large company that needed their feature and user base. Pivoting, spinning-off companies, creating white-labels and licensing features/benefits to other social media platforms and web sites can be lucrative paths for the right company.

MVP's are developed using a lean development process often referred to as 'agile' You will hear the terms lean and agile in your digital travels. Agile means that the customer and the software developers are working in tandem every step of the way. I have found great value and success in all my digital projects using that type of team approach, long before it had the cool 'agile' name.

During the agile development process, you will receive ideas and generate your own prototype, collect data, analyze it and learn. It is very much how we define the digital process throughout the book. Each iteration should be fast. It could be as short as a few days or a week. Review the results of each product iteration with customers and get their feedback. It is a partnership. Keep iterating until you have a fully functioning and disruptive product.

The key is IT and the business / customers must be partners. You are delivering a base product and then using feedback from the business and real customers to help you develop what people want, value and will pay for. There are no guessing games. The process works for B2B and B2C. I never had a software product or book that failed and I am confident I never will. The main reason is I listen to the customer and build the product they want and need.

Warning: As I mentioned at the top of this chapter I shy away from the 'Viable' historically used in building MVP's and much prefer 'Valuable'. Here is why: It is critical that you deliver value to the customer and not rush crap out the door and simply call it an MVP. You must deliver something that the customer will use. I am not saying to add all the bells and whistles upfront as that defeats the purpose of an MVP but if the customer feels it is not worthwhile they simply will not use it. So, you must strike a balance between underwhelming and overwhelming.

If you lose early adapters you will not have that all-important base you need to get feedback, adjust, publish the improved product and keep iterating until your customers are delighted and your organization is prospering.

Warning: If you build software in a glass-house you will likely fail. Glass-house hearkens back to the mainframe days when IT developed 'solutions' in isolation and people (often called end-users) had a 'take it or leave it' choice.
Unfortunately, the glass house mindset is not completely dead. Many vendors still waste time adding worthless bells and whistles when all they have to do is listen to their customers and build the product they want. I often get

demos of products and I am mystified why a company would add certain clearly worthless 'features'. Not listening to customers and not knowing the business are the reasons I came up with. They would save time, money and prosper if they only listened.

The companies that keep their eye on the ball are the ones that succeed in the long run. So, if you own or work for a software development company listen to what customers need. Let them drive your efforts. If you are buying software make sure you hook up with a vendor that listens to your needs and acts on your suggestions.

Warning: If you are creating a digital solution using agile, waterfall of any other planning process it is important you have a good way of testing the updated code before releasing it. Although, early customers realize they are not getting the finished product and want to help you build it, they will not tolerate careless bugs, misspellings, the same mistakes over-and-over, etc. There are too many alternatives for them, so do it right. You must also insure the new enhancements do not impact any existing features that the customer enjoys. I suggest a tiered software testing process consisting of 2 or 3 tiers.

3 tiers:
- Developer (Dev)
- Quality Assurance (QA)
- Production

Or 2 tiers:
- QA
- Prod

These are logical tiers rather than physical servers. Multiple ties can be done on the same server if they are partitioned correctly.

Warning: Do not migrate upgrades into the production environment without testing thoroughly first. This can cause pain for your customers, you and your team. I suggest you push back if someone suggests doing a quick and dirty change in production environment Friday evening before rushing out the door. Customers can be lost and your staff may have to work through the weekend to clean up the mess.

Warning for big companies that move very slowly: When working with MVP's test thoroughly but also promote software in an expedient manner. Too many sign-offs and wasted time going into production can destroy momentum and customer excitement.

How to Leverage Content, the Currency of the Digital World!

To be successful in the new digital world, you need valuable content and friction-less channels of distribution. Content is key to creating awareness and driving revenue. It is far more effective and cheaper than advertising. Content is the currency that just keeps giving and it is near and dear to my heart.

Every piece of content you develop is an asset. Organizations can benefit from developing and using content both externally to help clients and prospects and internally to help employees and students. Career seekers can benefit from publishing valuable content. It can attract potential employers and serves as the ultimate business card.

I have developed content for software products, social platforms, videos, audios and published 12 books (make that 13 including the one you are reading).

In this chapter I will share:
- What content is and why digital content rules
- Content types and channels you can benefit from
- Tips and techniques that work for me and can for you
- How to create and distribute valuable content
- How to easily repurpose content in many forms
- How to build and benefit from 'Extensible Content Objects'
- Why the Law of Reciprocity works

Throughout the chapter I will sprinkle many tips and techniques I have learned through experience.

So, what is content?
Content in its most basic form is information. Data, facts, figures, statistics, etc. are the foundational building blocks of great content, but there is more we must build on top of it to maximize its value.

Great content tells a story that people can understand and benefit from. It draws them in. It is all about them, not you. It goes beyond the raw data. It is about emotion and needs. It turns information into knowledge and knowledge into wisdom.

Digital content has a wonderful quality that although it costs time and energy to create, it has essentially zero cost to duplicate. Each copy, unlike the old analog VHS tapes or paper books, retains the quality of the original. I remember years ago, making copies of my popular Internet Marketing Video with VHS tapes. There was significant quality loss on the first generation copy and second generation copies from copies were grainy and difficult to watch.

These unique properties of digital content enable you to give it away for free if it fits your business model. Free is powerful. Free comes back to you in building your prospect list, web site traffic and the ultimate goal of sales.

Digital content can spread virally to millions of people around the globe in minutes by way of the Internet. It has happened to me. One example, which I describe in detail in the final chapter of the book, was the 'golden press release' I wrote that created tremendous traffic for GrandmaBetty.com. The avalanche of traffic temporarily brought our web server to a crawl. We received press coverage globally. That sparked word of mouth which further generated exponentially more traffic. We never needed any paid advertising after that one release.

Always make it about the person consuming your content!
If you make it your priority to help people you will be on the fast track to success. It is critical to clearly understand what your visitors, readers, watchers or listeners need and want. You will then fulfill their needs. Simply understand and alleviate their pain.

Work Backwards - why it pays to start at the end:
We discussed working backwards earlier regarding developing successful digital products and project plans. I use the same working backwards technique when writing a book, designing a webinar, teleconference, podcast or white paper. Working backwards works every time because to be successful:
- You must understand what the result looks like
- You must have a target
- You must understand their pain

Understanding what the end must look like will enable you to have clarity on the message you want to develop. You can then create a plan to get you there. If the content you are creating is relatively short, such as a blog post, newsletter/magazine article or press release, you can create a simple outline describing a beginning, middle and end. Then fill in the details.

If you are creating an eBook, print book, white paper, webinar or podcast your plan and outline will be longer but the basic steps will be the same.

- The beginning should state a problem the reader or attendee can relate to.
- The middle should describe how the reader or attendee can solve the problem
- The end should summarize your solution and in the case of a webinar or podcast also allow for audience participation in the form of questions and answers.
- You should also provide multiple ways people can contact you. An email address is a must and a phone number is important in my opinion.

What Not to Do!!
I recently attended a webinar hosted by one of the largest software companies in the world. I was looking forward to the webinar and agreed to take time out of my very busy work and writing schedule to attend, as the speaker's credentials were based in IoT and artificial intelligence. I was sure I would pick up some useful knowledge.

What I received was a 1 hour sales pitch on why I should buy their cloud product. They did not provide any valuable content I could use. I thought I was the only one that felt it was a waste of time, but two of my colleagues messaged me during the webinar, as they were equally disappointed. I will be very careful the next time that company 'pitches me' on attending one of their 'valuable' webinars.

What to Do!!
Always provide knowledge and wisdom. DO NOT make your content a sales pitch. If people think the only reason for your content is to sell them your product, they will be turned off. Bad content is worse than no content.

It is fine to describe your product or service at the beginning, end and where it provides value to the customer during a video, audio or written presentation. Do it in an informational way and not like a boardwalk salesperson who is trying to draw people into the game arcade.

For example, due to the success of my last book, 'The Ultimate Business Continuity Success Guide', I was invited to create white papers, speak on podcasts and lead a global webinar. I did not pitch my book, and to their credit, the sponsors I worked with did not turn any of the presentations into a sales call. During the global webinar (still accessible on the Disaster Recovery magazine web site) the magazine and the sponsor Everbridge.com kept it informational which made the webinar very successful. I was the featured speaker and lightly referred to my book only twice during a 52-minute talk.

Tip - Share your experiences and tell stories. People love to read stories and first-hand experience builds credibility and unique value. Don't hold back. Share, share, share!

Tip - Inform and educate. People love to improve. If you can help them improve you will be successful.

Tip - Write or speak with economy of words. Add value in every sentence. No fluff. People are busy. If you are respectful of their time they will appreciate it.

Tip - Be friendly. People like to work with nice people.

Tip - Be Original - Be Creative! Be Different! Be Yourself - use your own voice or that of your organization in every content piece. In all my material including this book I use my own voice. Now, I admit, that I can come off as being a 'little out there' but I must be true to myself to give you my best.

A confession: There have been press releases, white papers and even books I started to write without my 'voice'. They were bland and very business-like. I knew in my heart it was not me and I could not publish such material. In the end, I always used my voice and I always benefited from it.

Tip - Be Different! Lose the Herd Mentality: Following the herd or copying someone else's content is a sure way to turn people off.

People think that copying what worked before will lead to success. It does not work that way. If it was as easy as copying or patterning a book on other books, every book would be a best seller or maybe no books would be best sellers. Unfortunately, I see it too often and it does not work. What it does is create a crowded playing field where many books are competing for a limited number of readers.

If all your competitors are doing it one way they may be 'canceling themselves out'. Being different and going against the grain often separates you from the crowd and creates a competitor-free market of one.

People can be very bold though. I once developed a popular digital ecommerce software system that took me over a year to complete. One person who bought it thought it was a license to copy my commercial program and sell it under their name. That was unacceptable and sad. Unfortunately, in our digital world it is easy for people to CTRL C - CTRL V, in other words copy-and-paste. Never do it. It is a path to failure.

I have stopped using expensive enterprise digital software tools that contained written material that was word-for-word the same as other products. It was very dangerous and very bad.

Tip - Use Humor: I enjoy surprising and entertaining people by interjecting stories and original humor in my content, where you would never suspect humor would be accepted. The following is a great example:

When I published 'The Ultimate Business Continuity Success Guide' I did not expect to sell a lot of books and going in. I simply wanted to share my information to help business continuity, security, safety and technology professionals insure they could help their organizations survive any type of natural, human or cyber threat.

Many books on business continuity (over 1,600) are dry and humorless. I felt I could convey my experiences and mix in humor and some wild stories (as I tried to do with the book you are reading). For example, in a chapter on the importance of having an elevator speech I used a 'based on a true story', of a new business continuity professional taking a 'Twilight Zone' type of elevator ride with a VP of the company. When the VP asked the new person what he did, the new BC professional spewed out ridiculous acronyms only another business continuity professional would know. It meant nothing to the VP. The story became kind of funny and sad as the creaky old elevator climbed slowly toward the 72 floor and the VP was sweating and just wanted to get out of there. I injected humor but got my point across that you must be able to clearly communicate value defined by the person you are speaking with.

In another chapter I wrote a rather funny story of a company Vice President that demanded to lead a webinar that had an audience of top management attendees, although a capable person was set to do it. Well, the VP had his way and became the host. During the webinar, he became annoyed with one of the attendees asking too many questions and he started firing off instant messages to his colleagues calling the attendee a sh*t head this and that. He went on and on thinking only the people he was instant messaging could see his vile comments. Unfortunately, as he was the host of the webinar and he was sharing his screen so all the attendees saw every message. He 'retired to spend more time with his family' the next day.

The elevator story was fun to write and well received. In many of the reviews my humor was singled out as a positive.

One more example, when I was featured on that well received global webinar for DRJ Magazine on 'Digital Transformation for Business Continuity', I whispered to the audience near the beginning, 'Shhh, don't tell my boss but I love digital transformation so much that I would do my job for free, and if he is listening I just might have to prove it'. You can hear it online. By the way, he was listening and was the first person to call and congratulate me after the webinar. Also, I did receive my next paycheck.

Tip - Use emotion: Content that exudes emotion can be very effective. On occasion, you see it in a commercial or marketing piece in the United States. With the right digital product or service, it can be extremely effective.

Thailand, does it better than anyone. I have watched some of their digital commercials on YouTube over 100 times. They have a big impact on me and by the many millions of views they rack up I am not alone. They can teach you valuable life lessons. These micro-movies are as powerful, emotional and long-lasting as any full-length movie you will pay to see in the theater. The Wall Street Journal, Adweek and The Huffington Post are a few of the venues that have written feature articles on the Thai commercials phenomena. Here is a link to three of my favorite Thai commercials:

- https://www.youtube.com/watch?v=JPOVwKPMG8o
- https://www.youtube.com/watch?v=cZGghmwUcbQ
- https://www.youtube.com/watch?v=qZMX6H6YY1M

Please keep in mind the above Thai commercials were for a Life Insurance Company! What could potentially be bland and a good time to go to the bathroom, became must see TV! Humor or emotion can be used for most any product or service.

Repurposing Content:
Repurposing content means using it for more than one purpose. I have created content, such as webinars, that I have repurposed as the seed for white papers, social media posts, videos, blog posts and podcasts.

Recently I developed a webinar the was profitable. Hundreds of people globally attended it and it opened many opportunities for me. When I repurposed the webinar across other channels it became exponentially more profitable with little extra work. Through the years, I have repurposed content in every direction - webinars as podcasts, eBooks as webinars, blogs as special reports, etc.

Repurposing does NOT mean simply copying content from channel A to channel B. For example, the white paper I referred to would have been an awful webinar or podcast due to the density of material. But when used as the base and pared down to a few valuable points, I could effectively communicate in another channel it reduced my work by 85% and made for successful presentations.

Tip - If you will be repurposing material you create, make sure you retain the rights. I always copyright and indicate when creating content that it is non-exclusive to the vendor that asked me to do the initial presentation. If the vendor stipulates that I do not use the content anywhere else for 30 days I often agree, if it makes sense.

Extensible Content Objects (ECO's):

ECO's are a concept I developed from my programming background and they have served me well. ECO's go together with repurposing. They build another level of process, intelligence and building block 'snap-ability' into your growing empire.

When I started programming, I was weaned on procedural languages such as BASIC and Fortran. They worked well and were fun but it seemed that every subsequent program I wrote I was reinventing the wheel and thereby doing unnecessary work.

A different method of programming called OOP (object oriented programming) was invented in the late 1950's at MIT. It was not popular in the business world until the 1970's with Smalltalk and exploded in popularity in the mid 1990's with C++ and Visual Foxpro (I wrote 3 books on FoxPro back then - see I was good at finding trends).

Using OOP, computer programs were built using objects that could talk to each other rather than the procedure way of the past. OOP has many interesting concepts but a few I inherited (play on words as you will see) to the world of high performance content creation.

The Two OOP Concepts That Can Supercharge Your Content Creation Productivity:
1) Encapsulation - in simple terms this means an object binds data and the functions that manipulate the data. This makes the object safe from the outside world.
2) Inheritance - Ok, hang with me here because this is key - complex objects can be developed by combining simpler more generic objects (object composition).

For example, if you were developing a computer program / app focused on describing the features of types of cars (sports cars, SUV's, sedans...) you could save a great deal of time first building a simple generic 'car object'. That object would have the most basic properties of all cars: body, wheels, engine, wipers, etc.

You could then embed the basic 'car' object into a slightly high-level object 'sports cars'. Automatically all of the properties of the high-level object would be inherited and available in the more refined object - 'sports cars'

Now let's creatively apply encapsulation and inheritance to building your content empire. Earlier I described how I turned a webinar into a podcast and a white paper. The way I did it was to first create the webinar with basic information that could be used across all platforms - written, video and audio.

I saved the individual objects that made up the webinar and the webinar as a whole. I presented the webinar and it was a success.

I was then asked to do a podcast and whitepaper. I was easily able to inherited the base objects of the webinar into each new content format and simply customize and add details that took advantage of the unique properties of the new longer format environments.

Here is another recent success using content objects: in my 2017 book, 'The Ultimate Business Continuity Success Guide', I encapsulated hundreds of small content objects in a MySQL database rather than a word processing document. You can use a spreadsheet if you are more comfortable. The book had over 1,000 tips and techniques, so there were many little objects.

Having all the objects in a digital database in their barest form gave me great flexibility to extend, manage and organize them. When building chapters and sections of the 400-page book (more complex objects that could also be moved around) I simply used combination and inheritance to snap together the objects. It proved to be a simple task from what could have been a nightmare.

But that was only scratching the surface of the value the content objects provided: Maintaining the little content objects in the database separated the data layer from the presentation layer. This is very important and a very powerful concept which I describe in the next chapter, 'Why Data is the New Gold'.

It means you can easily present the data any way you want. Continuing with the example of the 'The Ultimate

Business Continuity Success Guide', beside using the content objects in the physical paper book and eBook formats, I was also able to create a valuable detailed database driven Online Business Continuity Roadmap with a few lines of html and JavaScript in seconds.

The roadmap became a very popular resource with business continuity, disaster recovery and cyber security professionals. Here is a link to the Roadmap if you are curious:

http://www.ultimatebusinesscontinuity.com/roadmap/

If I add anything additional content objects to the database, they will automatically display on The Roadmap for the next online visitor. The database can also be sorted in unlimited ways to create new value including apps. It is very simple and worth the effort to build this exponential value.

I hope you think about repurposing content and utilizing content objects to grease the content productivity skids. You can become a content producing machine. The concepts have proven valuable to me.

The Law of Reciprocity

The Law of Reciprocity – is a law I love in the physical and digital worlds. The essence is if you do something nice for someone they will feel compelled to reciprocate. The reasons for this are based in psychology and I have experienced it again and again. If someone does something nice for me I feel compelled to do something equal or nicer for them.

For example, a leading software company was developing an upgrade to their app. They sent me a $25, no questions asked, Starbucks Gift Card in return for a few minutes of my advice. Not only did I give them my initial thoughts on their product but I also had lunch with them and gave them many more suggestions on benefits that would appeal to a person like me. Whenever I see them at a conference I remember the lunch and the card.

Reversing the situation, I have sent autographed copies of my physical books to people I respect both domestically and overseas. I do not request anything in return. Some cannot believe I sent them an autographed book, especially overseas. Often, they write glowing, but honest, testimonials and reviews of my books and software. I then reciprocate with likes and shares of content they produce and a wonderfully virtuous cycle begins.

Unlike physical objects digital content has the unique quality of essentially zero cost after it is created. You can give it away for free and it costs you essentially zero. If your content has a high perceived value, there is a good chance great things will happen in return. Not with everyone, but you do not need everyone to become super successful.

I have successfully used each of the following content channels to build my brand and fill my pipeline for digital software, books and services. I hope you find them useful:

- Webinars - I have done 1,000+ webinars in my career. They are very cost effective and done right, can be extremely effective in building your pipeline. I could write a book on webinars and someday I just might do that. You can also host a webinar and have a featured guest / niche celebrity do the presentation. If you are doing the webinar you may be a bit nervous the first few times but the nervousness subsides quickly and soon you will be cool as a cucumber.
- White papers - these are fun to do and coveted by popular web sites that want to provide their visitors with detailed solutions. I recently did one on Digital Transformation for Business Continuity for Everbridge.com. I received great feedback although it was a bit longer than the typical 3 to 15-page white paper. I had a lot of information and experiences to share and took the time to share it.
- Apps - you can easily distribute your content as an app. Here are three methods to think about and possibly do further research if one interests you:

1. Responsive HTML5 is the most basic and simple way to create an app. Very little formatting is required to render correctly on different screen including desktop, tablet and mobile phone. Developing in HTML5 is fun and fast. It is within the ability of anyone that has light technical experience and the desire to learn a useful language.

2. Native apps are complicated and time consuming to create but offer the most functionality, if you need it. Unless you have extensive technical experience or a lot of time to learn I would leave this option to professional developers. You can find many on sites such as **Guru.com**.
3. A third category has proved very useful to me and my developer friends. I am often surprised how many organizations, including major software companies, who come to me for guidance do not know of it. Cross platform tools such as Titanium Appcelerator are relatively easy to use and offer access to many native mobile tools such as the camera and accelerometer. They are also cross platform so there are only minor changes to display content on IOS or Android. If you have moderate technical experience and the desire to build really slick apps you should consider this option.

- eBooks - if you have the time and the content a free or low priced eBook is very enticing to readers. Load it with valuable content and it will come back to you in the way of fans and possible new customers. When you write a book, people see you as an expert. Authoring a book builds credibility for you and your organization.
- Audio - re-purposing written content as audio is fast and easy. Listening to book is popular. Many people listen to books while multi-tasking.
- Video - is a very popular content channel. People that watch how-to videos on one of the popular video platforms such as YouTube may visit your web site for more information. You can demonstrate how your product or service works and connect with prospects. You can even create your own channel that people can subscribe to. My advice is you should make your videos information rich and not simply commercials.
- Blogs and websites - hosting a blog is an excellent way to generate interest in your product or service. A warning though, if you do not regularly update your blog it becomes a turnoff to visitors. Create a publishing schedule and stick with it. User generated content can help keep your blog fresh. Also, submit content you create to relevant blogs to reach a wider audience.

- Newsletters - writing and distributing a monthly newsletter is a great way to stay in front of prospects. Also, submitting articles to newsletters your target audience reads will broaden your reach. Insure that you get a byline including your email address and a link to your website.
- Seminars and speaking opportunities - I enjoy speaking and interacting with attendees. Depending on your product and the size of the audience it can be lucrative. You must weigh doing them against travel costs if the host is not paying.
- Create Infographics - these are appealing visual displays of data. I know social media platforms that release these and get instant visibility.
- RSS (Really Simple Syndication) - if you distribute content such as weather alerts or product news on a regular basis, this can be an easy hands-off approach. CNN and the NFL are examples of the millions of sites that leverage RSS. You can send headlines or more extended content. Users simply download an RSS reader and add a URL you provide them. Creating an RSS feed is simple and within the ability of anyone.
- Your Email Sig! - A sig is the signature at the end of your email. A sig typically includes a name, title, email address and phone contact numbers. You are most likely including that information already. Smart content marketers go a step further and wring additional value from their sig. It can be valuable real estate for your message. It is important to keep your message concise, to the point, interesting and powerful. Use action words rather than passive. You can include a line or two of valuable teasing content and a deep-link to a page on your web site (not your home page) if they want to learn more. Deep links pull more than a top link. There are many stories throughout the history of the Internet of companies that used their sig to make their message go 'viral'. For example, Hotmail had a sig that was instrumental in building their company into one of the hottest products on the Internet which eventually led to them being acquired by Microsoft for $500 million dollars.

- Editorial Letters and Features - Write letters to the editor and feature stories with valuable info and mention your name and web site. Most magazines and newspapers will print your site name. As an example, one of Grandma Betty's (she is profiled in the last chapter in the book) letters to the editor was published in Wired magazine and brought thousands of visitors to her web site. I have written many feature editorials, including my favorite on digitally hardening soft targets. Many found their way into blogs, newspapers and trade magazines and resulted in many interested emails and inquiries.
- Business Cards - The lowly business card is really a little billboard in someone's pocket. Don't cheap out. Have them professionally done. Put useful content on your cards so people will not throw them away. I missed so many opportunities to get potential buyers to my web site before increasing the usefulness of my business card. They can be important handy at networking events and parties. You never know when you will make a great connection. A value-laden business card is a great way keep your product or service fresh in people's minds.

Can a machine automatically create valuable content? You bet it can!

I have been following Narrative Science (narrativescience.com) for quite a while. They are big players in the AI space. I have met their team at conferences and witnessed demos of what they can do. For example, automatically transforming dry data in a baseball box score into an interesting story. I would not have known a machine wrote the story had I not seen it happen.

With the Internet of Things generating torrents of data, Narrative Science can be a great way to transform it into insights we can all understand.

They 'humanize data' with powerful natural language generation (NLG) and AI technology that interprets your data, then transforms it into insightful, natural language narratives with speed and scale.

They automatically turn thousands of data-points into actionable, valuable powerful stories and assets you can use to make better decisions, improve interactions with customers, and empower employees and students.

I hope you enjoyed this chapter on creating and leveraging content. Obviously, I am passionate about writing and sharing knowledge and it has worked for me.

Why Data is the New Gold!
How to Manage, Value and Protect It

As a digital company or professional you are in the data business. Data is your digital secret weapon. It is the modern gold. Understanding the power and value of data is critical to your business and your career.

Digital transformation, disruption and innovation are dependent on good data. IoT devices can suck in torrents of data which provide a new view of our world. Data is the nourishment for artificial intelligence, blockchain and other exponential technologies.

Data is information that can provide shockingly valuable insight and wisdom. I have used data to my advantage throughout my career. Insight can empower you to see the future through predictive analytics, make better decisions, get a decisive edge on competitors and create customer happiness. Throughout the book, I describe situations where people use data to develop products and services, disrupt industries and build fortunes. All the disruptors I mention in this book understand the value of data.

Most books on digital transformation do not dig into data on the level I will in this short section of the book. Fortunately, I have the technical background to do it. So please consider this a bonus chapter. I am a coder and was a technology project manager. I have written six books on database development that have been used by professional programmers worldwide, but do not worry this chapter is not meant to be a programming course and I am sure most of you would not want it to be. I simply want to pose some ideas, suggestions and general thoughts on managing and maximizing one of your most important assets - data. If you understand the basic concepts in this chapter if will be advantageous to your business and career.

Most companies I meet with, large and small, are paralyzed in some fashion by using the wrong data tools, siloed data and lack of data integrity. Unfortunately, they are devaluing their data. Bad data is worse than no data.

Toward the end of the chapter I will share some best practices for protecting data, insuring system availability and guarding laptops that often walk away. If you ignore these it can destroy your business. Data, system resilience and availability can be revenue builders.

Data silos must be eliminated.

Data must flow freely, horizontally and vertically in organizations. It cannot be siloed and hoarded. Data silos are a major reason organizations, large and small, do not get the value they deserve from the 'new gold' data. This has been an issue since the dawn of computers.

As a digital company, you will likely be collecting large amounts of data manually, using systems or with IoT devices. All data must be managed properly. Well managed data can lead directly to identifying opportunities and threats, greater revenue and reduced expenses.

Using the wrong tool or implementing the wrong vendor solution can result in a nightmare for you and your organization. It can become:

- Life threatening to your employees - in the case of stale data in a mass notification system
- Expensive for your organization where there is incorrect or duplicate data in an order system
- A frustrating drain on your limited resources and very difficult to reverse for you and your team

As a digital company, all sources of data are in scope including internal, external, structured and unstructured.

Structured data is data that fits neatly in rows and columns. For example, names, addresses, revenue and expenses.

Unstructured data can be conversations on social media, emails, text messages, etc. We can mine unstructured data for sentiment and what people are saying about your company. Remember, internal and external data are both valuable to you.

Managing Data - Database or Spreadsheet?

There are occasions where developing a simple database can be faster and much more valuable than building complex macros in a spreadsheet to make it 'almost' like a database. I will also point out use-cases where a spreadsheet is preferable to a database.

I will then conclude the chapter with some common database concepts and terms that will help level the playing field for you when speaking with your IT folks or vendors trying to sell you their expensive software solutions. When you ask them questions about fields, records, properties, data normalization, data models and API's, they will see you on a different level.

You will be in the top 1% of business users in terms of understanding the power of a database. The questions will help you select the best solution for your needs. It is critical to ask the right questions before you buy.

In fact, ask 10 of your non-IT business friends what the important concept of data normalization is or what a data model is and I will bet almost none of them will have any clue what you are talking about but these should be important to business people.

Spreadsheets:

Most professionals I know are more familiar with using spreadsheets than developing database solutions.

A spreadsheet is a paper-based or software program that captures data in rows and columns. Excel and Google Sheets are examples of popular spreadsheets. I have seen spreadsheets used for almost every application from financial to word processing to a home spun FTP(file transfer protocol) program using VBA (Visual BASIC for Applications). In fact, I have built some very complex spreadsheets using VBA. That said, it is important not to use a spreadsheet as a universal tool. Never use a hammer when a screwdriver is required.

I commonly use spreadsheets for simple contact lists, vendor features/benefits analysis and sometimes for more sophisticated reports than supplied by my business applications where I need to apply a further level of numeric analysis beyond what I can do in the core vendor tool. I also use spreadsheets for checklists and simple project plans. I find each of these use cases shine in a spreadsheet.

Spreadsheets can quickly get you into trouble when many people need to contribute or access information. You have probably seen spreadsheets emailed to 10, 20 or 50+ people for input. Each recipient enters information and returns it to the sender. It then becomes cumbersome and error prone to integrate and present the results. Keeping the spreadsheet up to date can be a time-sapping nightmare. The data in the spreadsheets rapidly loses its value.

At that point, if you are good with spreadsheets, you may start designing complex macros, programming in VBA or developing cross-worksheet formulas and pivot tables to produce metrics and summary results you need to do your job. Then, when you finally complete that effort, the need to produce additional metrics triggers another larger-than-necessary effort. Often this nightmare goes away using a database solution.

Spreadsheets also have the propensity to breed faster than rabbits in large and small companies. Siloed departments often have their own 'secret' version of 'the truth' or the 'not so gold copy' stored in spreadsheets out on their share drive, either on an official corporate drive or perhaps on their personal drive in the cloud, which is a huge security and business risk.

When spreadsheets breed and there are many versions of 'the truth', expensive trouble is in your future. Which spreadsheet do you believe? I have witnessed costly organization-wide data cleanup projects dedicated to trying to consolidate and 'normalize' this data, after the fact. It is a painful experience.

Databases to The Rescue

A database is a central repository that can manage vast amounts of data. If the database is designed properly gigabytes, terabytes, exabytes and even zettabytes of data can be sliced and diced any way you need, in fractions of a second. For example, a database consisting of millions or billions of records is trivial to manage and easy to derive important insights that can be the transforming difference for your organization.

A database is also an excellent tool when you need horizontal or vertical process input from many people throughout your organization. It is also valuable when you need to present the data in different ways to a variety of users with varying permissions. You can easily control what they can and cannot input and view.

Centralizing data has many advantages when mining the data and reporting on it. Merging information from many spreadsheets is not how you want to use your precious time. With spreadsheets, you might never see the entire story and worse you might see the wrong story.

Relational or Graph Database?

As of September 2018, relational databases are much more popular than graph databases, primarily because they have been in production far longer. Most, if not all, HR, sales accounting and inventory systems use relational databases.

As of September 2018, graph database systems make up only a small portion of the database universe. That is already changing as there are many interesting uses for graph databases. They allow us flexible and easy analysis of millions or even billions of data nodes. They integrate very nicely with artificial intelligence and IoT. Graph databases have been key to mapping DNA and discovering the causes of disease.

Many large social media sites, including Facebook, manage their user's information in graph databases. Because graph databases scale well, are flexible and beautifully visual, they help us discover critical well-hidden inter-relationships between entities in our organizations or any other facet of life. This can uncover opportunities and threats.

You can even graph past events and then reverse engineer them to discover cause and effect. For example, if a product or service goes viral and you want to understand the reason and then turn it into a repeatable process, a graph database loaded with prior news events, social media influencer's posts and perhaps some wonderful Google Trends (https://trends.google.com/trends/) information could uncover possible triggers! A relational database could not provide that level of interrelationships, nor is it built to do so. Always use the right tool for the job.

Basics of a Relational Database

It is important for you to have at least a high-level understanding of relational databases. It will differentiate you and will be very useful when building or buying solutions. Here we go - I will keep it interesting from a business standpoint and not too technical.

A relational database stores information (data) in one or more tables. Each table stores a common set of data in rows and columns. Employees, equipment, vendors, customers and applications should be stored in their own dedicated table, rather than one huge table.

A well-designed database solution separates the presentation layer (what the user sees) and the data layer (the back-end data tables). This separation gives you extraordinary flexibility and control in presenting and accessing data to users. For example, your back-end database can serve desktop and mobile users on any operating system or platform. It can feed big systems and small mobile apps equally well without changing the data.

Each table in a database is comprised of 'records' and 'fields'. Each field will store a data element which is a single piece of information. For example, in the employee table 'first name', 'last name' and 'employee ID' would be stored in individual fields. Employee ID, if unique, can be used as a special field called a primary key. Each table requires a primary key that defines the uniqueness of each record in that table. In some instances, a primary key can be the concatenation (combination) of more than one field. We can have two John Smith's in the database and differentiate them with the unique primary key employee ID.

Records: are comprised of groups of fields. That would include the three fields I just mentioned and more – city, state... in total to describe one person, which would constitute one record. For a Vendor record, fields might include: Vendor ID, Vendor Company Name and Vendor Representative.

If a database is properly designed, data about a person, a process or any other asset might be spread across multiple tables but it will be easy to bring everything together in reports and display information at a detail or summary level.

Proper database design is an important concept. If the relational database is treated as a 'flat file', sort of like building one big spreadsheet in the database, it will cause trouble down the road. Be careful as many people because of their familiarity with spreadsheets build databases to mimic spreadsheets. This mistake can easily become an expensive re-design and it will cause you lots of sleepless nights thinking about how to fix the mess. Your database MUST use the power of proper powerful relational design. It is not complex to do it. When you go beyond simple databases you might want a database professional or power use to design your database.

Normalization – is the process of organizing the columns and tables to reduce data redundancy and improve data integrity. Normalization occurs by breaking out repeating data into related tables. In my experience, there are instances in database design where partial normalization makes more sense from a response perspective.

If your database is not normalized, the reporting demands will quickly turn into 'kludgy' (a tech term for messy) solutions in an attempt to produce quality output. Worst case is, you can wind up with a very expensive cleanup and database re-design on your hands. Trust me, you DO NOT want to go there.

The best solution is for the database to be designed properly upfront and the data should be automatically imported if possible or manually entered in the system abiding by strict rules for consistency.

If you are fortunate enough to have corporate central 'gold sources' of data within your company with unique identifiers (primary keys) you may want to upload that data to your database. You can even set up automated SFTP (secure file transfer protocol) uploads to add, modify or delete data automatically as often as required. I often do this with large dynamic SAP, PeopleSoft HR databases and asset inventory systems.

The bottom line is databases might sound complicated but they are not. Just take it slow and build your skills at your own pace. Unless you are a database professional, no one expects you to develop complex databases. Knowing when to use a database as opposed to using a spreadsheet or another tool is the key and very important to your business and career success.

Data Protection and Cyber Security
This is not a book dedicated to cyber security, data privacy, disaster recovery or business continuity but I would be remiss to not include some thoughts as I have expertise in those critically important digital fields. Every company is now a technology company and data is your second most valuable asset - people are always first in my mind.

Data breaches are front page news and you do not want to be on the front page for the wrong reason. I implore you to take the protection of data and systems very seriously. If you do not have expertise in-house, it may be prudent for you to contract with a cyber security or business continuity/disaster recovery company to examine your infrastructure, policies, procedures and perhaps do system penetration (pen) tests to expose and mitigate vulnerabilities such as cyber from the outside and inside, phishing attacks, spoofing, malware, DDoS (distributed denial of service). I have witnessed too many preventable breaches in the last year to companies you would think would have their house in order.

In 2018 the European GDPR (General Data Protection Regulation) took effect. These are serious data regulations with high penalties. Even if your company is not located in Europe but you do business with European citizens you are in scope for these regulations. In addition, in the United States New York has already is one of the first states to implement stringent data protection / privacy regulations and I guaranty more states will follow probably by the time you are reading this.

When I interview C level executives and entrepreneurs and ask, 'what keeps you up at night', data, cyber issues and unavailability of systems is at the top of their list right after safety issues and they are right to be concerned.

Cyber from the inside, is as much or more of a problem than it is from the outside. Only give people access to what they really need. System non-availability keeps everyone up at night, up to the C or founder level. Often, we cannot go back to our manual systems the way we did prior to our dependencies on computers.

Data leakage is rampant. I will not go into detail but trust me it is scary what is mistakenly uploaded (or maybe not mistakenly) to the public Internet. Be careful where your data ends up.

Be careful about employees having access to personal cloud drives at work. They may be storing sensitive data on them. If the cloud drive vendor is hacked, or an employee leaves an organization, you may have a very serious and very expensive problem on your hands, and you don't need that.

Malware and ransomware can have serious implications. WannaCry impacted some of the largest companies in the world including banks and hospitals. Always patch your systems as quickly as possible.

Limit open ports to the Internet to only the essentials. Open ports are doorways into your infrastructure for hackers.

Make sure you have redundancy at every point upstream and downstream for every critical system. You must do system failover tests on a regular basis to insure you can recover your system in the time-frame your business requires. You do not want any surprises during a real event.

Laptops Have Legs - Do Not Let Your Data 'Walk'

The fact that modern laptops are small, light and mobile means they can also be high-risk for us and low-hanging-fruit for criminals. That is a dangerous combination. Whether the criminal's intent is to resell the hardware or sensitive information stored on the laptop the impact to you and your company can be extremely high. One stolen laptop with sensitive data can destroy your business.

Here is a true story that hopefully does not remind you of your company: I learned of a company that left laptops unlocked and un-encrypted on cubicle desks near exit doors. A thief shoulder-surfed behind an employee and easily gained access to the office. Supposedly it was a secure work area but unless you are using a turnstile or there is an unusually alert $12 an hour guard on duty, you know as well as I, not everyone badges-in. Few employees entering a building question the person behind them, especially in mid and large companies where you only know a small percentage of the employees. This is way too easy an opportunity for the bad guys to steal your sensitive digital assets!

The shoulder-surfing thief easily got access to the office, grabbed six laptops, threw them in a backpack and scooted right out the door. Sure, the entrance and exit doors were under video surveillance but by the time the video was reviewed the thief, laptops and more importantly their entire unencrypted customer database was long gone. Probably the data was already sold to a competitor or on the dark web being sold to all takers for pennies on the dollar by the time the theft was realized.

In another true story an executive placed his laptop under his seat in an airport waiting area for 'just a couple of minutes' to go to the bathroom. The laptop might as well have had a sign on it 'Please Take Me'. Well you guessed it, when he got back from his 'business' his laptop was gone-baby-gone. The executive was shocked and panicked. I could list hundreds of similar stories I have knowledge of and I am confident you could as well.

Tips to help protect your laptops AND especially your data:
Tip - NEVER SAVE UNENCRYPTED SENSITIVE DATA ON A LAPTOP OR USB DRIVE! Sensitive data should reside within the walls of your secure network.

Tip – Create a laptop security policy and publish it to all employees.

Tip – Laptops should have remote tracking devices activated. Depending on your employee/union environment this may be a challenge to implement. Partner with HR and legal, if necessary. If you are a startup this will be less complex to initiate.

Tip – Physically secure your laptops in the office. The story I described above is but one of many of criminal's shoulder-surfing or otherwise social engineering their way into lightly secured workplaces and stealing laptops. Remember, it only takes one laptop with sensitive customer or employee data to put your company in the headlines, for all the wrong reasons.

Can your office environment be compromised? Think about it. No, better than thinking about it take a walk around your office, factory or warehouse. Shoulder-surf in behind people you do not know. Let management know that you are going to perform a 'laptop theft scenario' in advance so you do not get in trouble.

Place a test laptop in a cubicle and then come back and walk out with it. Did you meet with resistance or did the security guard that does not know you hold the door open for you? If you were stopped and questioned, great! If not, fix this risk asap.

If there is even a remote possibility that your laptops and data can 'walk', fix this risk today.

Losing $5k-$10k of laptop hardware is bad but possibly losing sensitive data can be devastating. Fines can be in the millions of dollars and high level C heads will roll.

Tip – Desktop and laptop USB ports should not accept unauthorized USB drives! They can be disabled or programmed to only accept authorized devices. There are many horror stories of people using USB drives to plant malware and viruses. One shiny new USB drive picked up in a parking lot by an unsuspecting employee and popped into a networked laptop USB port can bring down a network, after sensitive customer data has been siphoned off. Believe it or not, this happened to one of the top data security companies in the world and it impacted many of their global clients, including some of the largest companies in the world. All from one evil USB found in a parking lot by an HR employee and innocently popped into a networked laptop. Once a network is compromised it is difficult to insure it is completely void of malware 'sleeping in the background'.

Tip – Laptops should ALWAYS be kept close-at-hand and attended to when traveling. Let me stress that – ALWAYS!

Tip – Laptops should never be left in the trunk of a parked car. Cars do get stolen and criminals can follow you after seeing you deposit a laptop in your car trunk. Trunks are easy to break into in seconds. Blink, and your lap top and data are gone!

Coincidentally, I had written the above tip a couple of weeks prior to a government laptop with sensitive data being stolen from an agent's car trunk. I wish I published this chapter earlier and the agent had read my book. Maybe the agent would not have stored sensitive data on his laptop and would have left his laptop out in the trunk.

How to Buy Digital Technology, Be Happy and Never Get Burned!

You are building a digital business and career. You need great software tools and possibly IoT devices to succeed. The right tools are like magic but implementing the wrong tools will cause long lasting pain and can derail careers. Sometimes the wrong tools are bought and even never get implemented. On other occasions, they are bought, rolled out to users and then must be replaced with a better tool and users have to be retrained. Ouch!

I have many years of experience evaluating and implementing software in enterprise size corporations and startups. It is what I love to do. I was the person working for a leading global financial institution that traveled to Las Vegas, Orlando and other not so glamorous locations to discover digital tools and technologies that could make a difference in our business. I would evaluate interesting technologies and if they proved to do what they said they could do, I would implement them for our developers and/or business users. I had to get it right, every time and happily I did.

This streamlined chapter on best practices when researching and buying digital technology can save you time, money, stress, anguish, finger-pointing and best of all it can bring your customer and you delight.

Software and IoT Tool Selection Tips:
Define your requirements first to ensure the product you are considering meets your needs.
- Focus on benefits and not features. Do not worry about meaningless bells-and-whistles that will not benefit you, your employees or your customers.
- Do your research. It is easy to find in-depth reviews and articles, both online and offline.
- Seek stability. Your digital tools must be available when you need them.
- Seek ease of use. If tools are not super easy and intuitive, people will not use them. They use simple tools like TurboTax at home. They are used to Amazon One Click. Allow users to sit in the driver's seat when you are evaluating products.

- Seek integration. The ability to integrate with other systems can get you to world-class. **ProgrammableWeb.com** has 19,949 APIs that enable you to mix and match products.
- Check vendor references, but do not stop there. They may cherry pick some of the references they provide. You should dig deeper. Seek good and bad references and reviews. You must make an educated decision.
- Decide if it makes sense to buy, build, or use a free solution. In my experience with enterprise systems, it makes sense to buy before build.
- The most valuable suggestion I can share is if you are buying a critical technology tool that can improve internal processes or provide customer delight, research products, determined the finalists and…

Always Do a Pilot

You MUST 'kick the tires' and play with all facets of the system BEFORE signing on the dotted line!

A pilot is a free trial of the software you intend to roll out to your organization. If the vendor insists on a small fee to do a 'proof of concept' it may be acceptable especially if you can apply the fee toward the purchase price if you decide to move forward.

I have purchased many enterprise systems and I have written and sold enterprise software solutions to some of the largest companies, educational institutions and governments in the world. I have a perspective from both sides of the fence:

Tip – Even if you have done a thorough product/vendor feature, benefit, cost analysis and you have attended the 'dog-and-pony' demo by sales and pre-sales engineers and seemingly received the answers you were looking for in your request for proposal (RFP) – it is simply not enough. That is the starting point. Now is the time to do the pilot!

Tip –You must also stress-test the system. Trust me, poorly designed systems can behave VERY differently with 200 records than with 20,000 or 200,000 records. The last thing you need is a system that was snappy and peppy when the vendor demonstrated it, but in production it became unacceptably slow. Users will not tolerate latency; 'this is not the 1990's World Wide Wait'. You do not need that aggravation.

Tip – A pilot is the ONLY way you can truly be sure the system is right for you! I really want you to do this for your own sake and that of your company.

Tip – During a pilot you will see the imperfections that the salespeople conveniently forgot to mention during the dog-and-pony demo. Be especially careful when it is near the end of the quarter and the sales person puts on the hard-sell to make his/her quota or bonus. Think before biting on their pitch.

Tip – Ask your IT team to attend any of the technical meetings with the vendor. They will hear things and see red flags that you might not. Get their blessing on the technical aspects of the system before signing the contract.

Tip – I suggest you have a few users on different levels of technical comfort/discomfort try out the new system. Don't only have power users. Have 'newbies' as well. They will give you great feedback, both good and bad. Speak with them and survey them:
- Do they like the user experience (UX) and interface (UI)?
- Do they like the reports produced by the system?
- Is the system overly complex?
- Is it slow?

Then, buy them lunch or give them a cool t-shirt for testing the system.

Any quality vendor will be happy to help you do a pilot or free trial. They want to insure you are getting what you expect. Otherwise, it will bad for you and bad them down the road. It will not end well if you are both not on the same page. Please accept my advice on this one.

Tip for Vendors – As a I feel good when a vendor strongly advises I do a pilot. It shows me your product will speak for itself. Words are cheap – 'the proof is in the pudding.' You should encourage prospects to do a pilot. If your product is great you will make more sales with a try before you buy offer. You will also avoid customers that had expectations of benefits your product does not provide. You do not need angry customers.

The bottom line is, do a pilot EVERY TIME you are considering a new system. You will save yourself time, money and lots of frustration.

How Two Technologists are Connecting Programmers Worldwide

devRant (devRant.com) is a unique, rapidly growing online community focused on the lifestyle needs of software programmers. Edgy and fun, it's not your typical tech community.

I visited devRant.com and picked off some representative posts, out of thousands, that you do not usually see on social media sites:

'I just discovered devRant today and got everyone in my office to downloaded it. A couple of them made me laugh so hard I was almost crying. They were so relatable it was scary.'

- 'I f*!@king love devRant. All the support here seeing the new upvotes every time I open the app makes me smile and just remember why I love making stuff.'
- 'I got addicted to devRant and replaced my Facebook browsing with it. I really enjoy the stuff posted over there and posting things myself. It's a site every developer should read and be active on.'
- 'I've spent the past 3 days sitting in front of my computer learning or actively programming, with occasional dips into devRant reading your stories, frustrations and victories and I truly feel at home.'

devRant was built by two digital technology innovators. It launched March 2016 and has grown exponentially month-over-month with zero advertising. In this chapter, they share their innovation and digital technology secrets of success.

Co-founders, David Fox, Director of Engineering, and Tim Rogus, Director of Design met at a multi-media development company where they developed a platform for one of the largest and most popular dating apps in the world. The system they built consisted of over 5 million members. Using a graph database instead of a relational database mixed with creative and powerful algorithms allowed them to achieve functionality beyond that of even Facebook. Their success was highlighted in mainstream news outlets.

What makes Rogus and Fox different from some entrepreneurs I meet is their relentless approach to customer delight. They never throw features out there randomly and 'see what sticks'. That sort of scatter-shot approach turns people off. Instead, both the founders actively use the devRant platform and listen to their visitors. This allows them to provide new enhancements users want, instead of what the founders think users want. It is an iterative process and feedback from software developers overwhelmingly indicates their delight that improvements to the platform are user driven.

I interviewed Fox and Rogus at a coffee shop in Manhattan to learn why and how they took on such a big challenge and the secrets of their success.

So, why devRant?
Like all devs we have workplace gripes. Sometimes we just want to vent and communicate with our tribe. We could not find a programmer lifestyle app or platform so we built one. If you need something, probably other people do as well.

Tell me about the early days of devRant:
We launched devRant March 2016 as a Minimal Viable Product (MVP). It was a simple to use IOS app with a clean visually attractive user interface (UI) and a strong back-end.

There clearly was a need, as we had hypothesized. The first month we had to get our feet under us and we had a smattering of first movers using our app. Month two the floodgates opened and member growth has steadily increased month-over-month.

Many of our members now spend more hours actively participating on devRant than they do on Facebook, Twitter, Stack Overflow or Google. In fact, membership on some general social media sites such as Facebook has trended down by double digits while our targeted membership is growing rapidly and virally. We currently have members located in 100+ countries and on every continent except Antarctica. Hey, if you happen to know programmers in Antarctica please tell them about devRant.

Why do you think programmers are so passionate about devRant?

We immediately hit a chord with programmers that were looking for a place they could talk about anything with their peers. Programmers call devRant their home. We stay true to our members rather than trying to please people outside our target audience. Anything goes on devRant as long as people respect each other. Amazingly it is the respect and good nature attitude of our platform that people find unique. They feel so comfortable. We have each other's backs. Many members came to devRant from other platforms where they were blasted when they posted questions or their opinions. Attitude is important to people.

We listen to our members every step of the way. It is all about them. They are our partners. Essentially, they built devRant. Sure, we built the technical infrastructure but we made sure we were building what they wanted, not what we thought they wanted. Always listen to your users. We used surveys, questions and gathered lots of ideas for improvement from member's posts.

How did you grow your user base?

We have grown organically. Often, a programmer will find out about devRant from a friend or a post finds its way to the search engines. When someone starts using devRant in a company or school it spreads virally. If you build a product people need and want word spreads fast.

Our Android app is recommended on the Google Play Store and brings in a tremendous number of users. IOS also does well and our revamped web site is getting great feedback.

devRant has been featured on the front page of Reddit, Forbes.com, SDTimes.com, The Next Web, Fossbytes and many enterprise software sites. We do guest posts for software companies such as Appcelerator (cross platform development tool) and Neo4J (graph database). These posts are value driven - not advertisements.

In some cases, we simply generate an interesting infographic from our rapidly growing data repository and it is instantly published by well-known news sites seeking to provide unique content to their readers.

We both speak at selected innovation and technology conferences. Our ideas are often published on major enterprise software company blogs and emags. Startups and large companies come to us for guidance and we are always interested in sharing. That is one reason we post our technology stack on our web site. These initiatives combine to drive significant focused traffic.

I am fascinated by your swag. I saw it mentioned in thousands of posts.

To help spread the word of devRant we let users earn free laptop stickers (developers love laptop stickers!) which we knew would end up on laptops at work, school, meet-ups and conferences. We have shipped devRant stickers to thousands of members globally. Our members also love the cute squishy devRant stress balls awarded to posts that meet an upvote point threshold on a single rant.

Did you spend a lot of money on advertising?

We never needed advertising. We did not spend one cent on it. If you create a product people need and want and position it right you do not have to spend a lot or in some cases anything on advertising.

How big is the programmer market?

There are currently 18.7 million software developers world-wide. This is estimated to grow to 27 million by 2021. As new exponential technologies come to market, the need for programmers will grow as well, which will increase our user base.

Who is the competition?

We have no real competition for our niche. There are platforms where you can post a technical question and get answers. That is not our niche. We like to stay focused. Many programmers that use those sites also use devRant daily.

Can you share how you understand what sort of enhancements people need and want?

Sure. We listen to our users. Tim runs our product analysis efforts. We use tools such as Amplitude for event and user tracking, as well as Google Analytics for web. Sometimes we pull data from the database and we run an analysis in Excel. We have an A/B testing framework in place that has allowed us to optimize our algorithms as well as add more visual features.

Can you give readers an example of enhancements driven by member requests?

The devRant podcast series features industry leading guests. They are 'delighted' (in their words) to participate and talk to our community. We were pleasantly surprised at the tech icons that want to be featured guests. The people we ask to participate always say, 'sure, let's do it'. Next thing you know we are interviewing them and they are all great fun, informative guests and nice people.

We had tremendous demand for an API from the community so they could extend devRant. Our community consists of thousands of developers that love building things. The devRant community has created many awesome projects using the devRant API. We help publicize their projects.

We had requests for avatars. Tim designed some amazing avatars that can be customized in detail based on the level of points members have earned. We add new avatar accessories on a regular basis to keep it fresh and exciting.

We had requests for a cartoon series and Tim created some fun ones that people love.

We had requests for the ability to follow your favorite members and their posts. David built it and the community was delighted.

Tim also created a devRant comic book which has become very popular.

There are many other popular member requested enhancements we have implemented. We are currently finalizing others that will provide significant value to members.

I assume as an early stage start-up you do not generate any revenue. Is that true?

Fortunately, that is a false assumption. Although we never stressed monetizing devRant the developer community insisted on it. They are committed to devRant's long-term success.

We have a popular store consisting of devRant branded products. Our t-shirts, sweatshirts, hats, mouse-pads and growing product line is on every continent. The reason each product sells so well is the community asks for them and we comply or if we come up with an idea we get their buy-in before launching. A great example is our spin-off site named devDucks.com.

Ducks are popular in the programming world. The exact term "rubber duck debugging" has 61.900 results and 'rubber duck debug" has 14.300 on Google. Andy Hunt, a programming icon and one of the 17 original authors of The Agile Manifesto, made Rubber Duck Debugging famous. He discussed the idea in his landmark book,' The Pragmatic Programmer'. The idea is, programmers explain coding problems to their duck and the duck helps them solve the problems. Believe it or not, this type of interaction has been scientifically confirmed to have real-world value when solving problems.

So, we created a site devoted to ducks for developers. Programmers can purchase a duck or a flock of ducks and customize them with 20 types of programming language capes. Some programmers have purchased flocks of ducks and all the capes. Our devDucks have waddled to all corners of the earth. Oh, by the way Andy Hunt was a featured guest on the devRant podcast and he was incredible!

devRant will always be free but we offer a premium monthly membership level with lots of devRant goodies for a small fee. It was requested by our user base and it is very popular. Many users have become members.

What technology do you use on devRant?
We are focused on ensuring devRant always provides a great customer experience and that means using the right technology tools. We get many nice comments from users who benefit from our elegant easy-to-use interface, powerful elastic search engine, configurable instant notifications and a highly relevant algorithm that learns a users' behavioral preferences and presents relevant rants.

We openly share information about the tools we use, as sharing is what devRant is all about. Sometimes that surprises visitors. On occasion a new person will come to devRant and try to learn from current users what technologies we use to create, power and monitor devRant.

They just send them to our web site home page devRant.com where everything is listed.

What's next for devRant?
That is up to our members. The precise enhancements will depend on what they want and need. We have been approached by potential complimentary services to help programmers with career services and learning opportunities. Our members always come first and we will partner with them to decide what provides optimal value.

People enjoy meeting in the physical world and our members have organized devRant meetups on many continents. We want to help formalize that in the future.

We are also constantly testing new tools and exponential technologies that can provide benefits for our users. Technology progresses at a fast rate so you can never sit on your laurels too long. We are passionate about technology and pride ourselves on having a world-class technology infrastructure.

Final question. I read a lot of complimentary posts on devRant. You guys have become heroes building a home for programmers to share whatever is on their minds. Does it ever get to your heads being tech rock stars?

(Both guys smiling) - Our members are the heroes, but yeah, sometimes we pinch ourselves.

You can learn more about devRant by visiting devRant.com or by downloading the devRant Android or IOS apps.

How an Innovative Recycling Company is Changing the World

In nature, waste does not exist. All materials are reused or recycled through natural processes. Humans have created complex plastic polymers and technology hardware that results in waste. We have broken the closed-loop, sustainable system that has existed on Planet Earth for billions of years. The problem is growing with the nonstop demand for safe, conveniently packaged consumer goods and hardware such as mobile devices, laptops and billions of IOT devices. This is annually creating billions of tons of non-recyclable or difficult to recycle waste.

TerraCycle, is the ambitious dream of Tom Szaky. He founded the company in 2001 as a college freshman. The goal was to manufacture a simple organic fertilizer. He and a fellow student fed leftovers from their cafeteria to an army of worms, then liquefied the worm compost into a completely organic, ultra-effective fertilizer. They did not have much money and could not buy the packaging they needed to start selling their fertilizer so they bottled it in used soda bottles they collected from recycling bins, unwittingly creating the world's first product made from AND packaged entirely in waste. Innovation at its best!

Tom, a true entrepreneur, emptied his savings accounts, borrowed money from friends and family, and maxed out his credit cards to create a massive worm poop conversion unit. Most of Tom's time was spent shoveling rotting food out of the back of Princeton University's cafeterias. Broke, exhausted, and ready to throw in the towel, Tom met Suman Sinha, an angel investor who cut the young entrepreneur a check and became TerraCycle's first investor. With the money invested by Suman, Tom could rent his first office space in New Jersey.

TerraCycle grew through the years and is now an international leader in recycling what was previously not recyclable. TerraCycle's founders realized the revolutionary idea they discovered was not worm poop, but using waste materials which have no value to make products that are innovative and affordable. I believe the preceding sentence is so important for two reasons. 1) The ability to make something useful from garbage is transformational and 2) Thinking broadly about the real problem and providing the most customer value. In this case the value is to every human on earth. I especially like the way they saw the 'big picture' of how they could help the world.

TerraCycle has come a long way in a relatively short time. They partner with major consumer goods manufacturers such as Procter & Gamble, Bausch + Lomb, Colgate-Palmolive, Tom's of Maine, L'Oreal and many more to run free collection programs. More than a billion pieces of pre and post-consumer packaging have been collected and over $21.5 million has been donated to schools and non-profits. The collected material is turned into a raw material that is sold to manufacturers to use in new products.

These recycling programs have been running since 2007, when the first was launched with organic beverage manufacturer, Honest Tea. In 2009, TerraCycle took its successful recycling programs overseas and today operates in 21 countries, including the UK, Brazil, France, Germany, China, Japan, Canada, Mexico and Australia.

In 2014, TerraCycle launched Zero Waste Boxes which allow individuals or groups to recycle virtually anything in their home or business, other than hazardous materials. The collection boxes, ranging from automotive parts to cooking oil, are sold on TerraCycle's website and through a variety of online retailers.

TerraCycle and Tom have received numerous social, environmental and business awards and recognition from a range of organizations. In 2006, Inc. Magazine named TerraCycle "The Coolest Little StartUp in America" and the #1 CEO in America Under 30. More recently, in 2013, Tom was named a Schwab Foundation Social Entrepreneur of the Year and a Forbes Magazine Future 30. In 2016, Fortune magazine listed TerraCycle as one of "7 World-Changing Companies to Watch." In 2017, TerraCycle was a winner of the United Nations Momentum for Change award and recognized by the U.S. Chamber of Commerce for its programs.

I had the good fortune to interview Ernie Simpson, the Global VP of Research and Development for TerraCycle. Ernie shared valuable information regarding innovation and using new technology to continue making TerraCycle a leader in sustainability.

Ernie previously had a successful career in research & development for Johnson & Johnson and DuPont. Soon after retiring he decided to seek an exciting new challenge. He saw an ad on Indeed with TerraCycle. He went on the interview and it was TerraCycle's lucky day. Tom and Ernie had the common mindset to be as innovative as possible in the right way. Up until then the recycling industry had an image that needed improvement. Ernie brought methods for specification and proper process that raised the bar and improved the image of the recycling industry.

Eight years later Ernie and TerraCycle have taken on many big challenges and consistently won. Here are three of the many ways they are helping our people and planet:

Diaper recycling - 6% of landfills are basically dirty diapers. TerraCycle has developed a method to significantly reduce that and already has a client in Amsterdam using their solution.

Cigarette butt recycling - the US National Institutes of Health (NIH) estimates that approximately 4.5 trillion of the 6 trillion cigarettes smoked every year are littered in the environment. Shorelines, sidewalks and parks are degraded with unsightly and unhealthy butts. The waste collected through TerraCycle is recycled into a variety of industrial products, such as plastic pallets, and any remaining tobacco is recycled as compost.

Ocean Plastics - when I learned from Ernie that 25% of the worlds plastics end up in our oceans it had a life-changing effect on me. Immediately, I rethought a 21st century digital message in a bottle I had created with the intention of sailing it across the Atlantic Ocean. I quickly realized that what I was attempting could add to the waste problem.

TerraCycle has vigorously attacked the ocean plastic waste problem head-on and I am happy to report that they were honored with the United Nations Momentum for Change Lighthouse Activity award for recycling ocean plastic into shampoo bottles. The program is growing and you can help and read more about it on their web site.

Ernie is a life-long innovator. He always wanted to take on complex problems and simplify them. He strives to develop eloquent and precise solutions. Eloquent is a term great computer programmers also use when they write simple yet powerful solutions. When I was a professional programmer my code was 'almost eloquent'.

Ernie is a problem solver. He says when he knows the cause of a problem he can build a solution. If someone does not quite know the cause but describes the effects Ernie can still peel the onion, find the root cause and build the solution.

Ernie advises people to seek challenges with a positive attitude. The word "can't" is not in his or my vocabulary. He embraces challenge in his work. It sparks innovation. Mundane, boring jobs can be tedious and people lose their way.

Ernie, like, Neil deGrasse Tyson, the late Carl Sagan and Stephen Hawking has a knack for describing complex problems in simple terms. During our conversation, Ernie explained complicated processes in a nice straightforward way, that even I could easily understand.

Ernie described how technology plays an important role for TerraCycle. It vastly improves loading using optical sorting and near infra-red technologies. These techniques enable sorting by shape and color.

Ernie envisions exponential technologies such as robotics playing a role in sustainability in the near-future. Robots can tirelessly go into piles and intelligently separate various types of materials into compost.

Ernie also described a valuable new service TerraCycle is pioneering that can potentially change the packaging industry. By using the knowledge they have built through years of separating materials and rebuilding them into useful products, TerraCycle can help clients design and build lighter, less complex packaging at a lower cost and with greater value. For example, a juice box on a shelf may absorb ammonia from a cleaning fluid being applied to store floors. When that occurs, the juice can taste bad and be harmful. TerraCycle can suggest improvements to the designer to make the packaging more resistant to harmful chemicals.

Ernie pointed out that for various reasons only 25% of waste is recycled. People are hesitant or they may not be aware of how to recycle. TerraCycle has a near time goal to double that to 50% recycled. TerraCycle will innovate as much as they can to make it happen. Having spoken with Ernie and having read Tom's book, I am confident they will do it!

There is a wealth of information on recycling and how people can help make our world a better place on TerraCycle.com. I know you will find it valuable.

How a 185-Year-Old Railroad is Creating Digital Delight

The Long Island Railroad (LIRR) operates from Montauk Long Island to final destinations at Penn Station in Manhattan and Atlantic Avenue in Brooklyn. It is the busiest commuter railroad in North America transporting 350,000+ riders daily. The railroad operates 11 branches and 124 stations over 700+ miles of track. It originated in 1834.

I have a mostly favorable relationship with the railroad. I benefit from using the railroad for most of my daily commutes to work. I save on wear and tear on my car and help to reduce pollution.
There are more delays than one might hope but I 'make lemons into lemonade' and use the 'extra' time to write a large portion of my thirteen books, including the one you are reading. My 2017 Amazon best-selling book, The Ultimate Business Continuity Guide: How to Build Real-World Resilience and Unleash Exciting New Value Streams', was 396 pages. I wrote 90% of that one during my commute, so thank you LIRR.

Unfortunately, many riders do not feel as I do and I understand their frustration. The overwhelming majority of morning and evening rush-hour riders are business people or students that must be to their destinations on-time. The LIRR has often been beleaguered with delays, mechanical problems and signal issues. It also has a history of poor communication to riders during delays or cancellations. Unfortunately, during a disruptive event communication is very important. In my career as a Business Continuity Director for large company's good proactive communication during crisis events has proved extraordinarily valuable to people impacted by the event.

Happily, the railroad is beginning to make progress to digitally delight their riders, well at least me.
For example: One morning during March 2018 I woke up at 5:45 AM to catch an early train to work. I like to get into work early when I am writing a book. It enables me to comfortably write on a sparsely crowded train with no sitting

next to me. I also can put in 30 minutes when I arrive at work and still begin working before my scheduled start time.

Before I left my house, I received a text message from the LIRR that there was a serious delay due to an event beyond their control. Pushing this information to riders is important rather than having to visit a web site and pull the information.

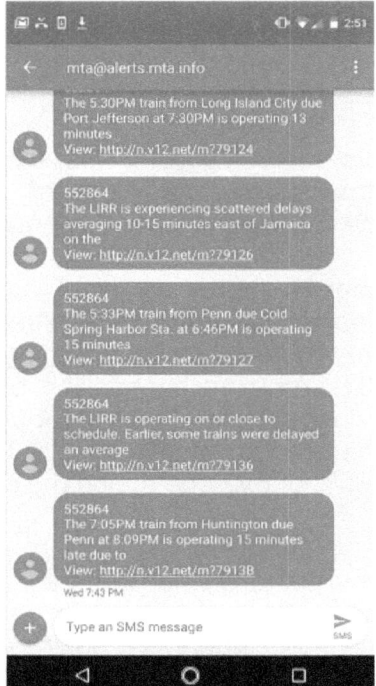

After I received the text alert I checked the LIRR TrainTime app for more detailed information. On the main page of the app I was informed that there was a serious incident with a passenger on the tracks on my branch. The app clearly explained what happened and which trains were canceled or delayed. They also used their incident database and shared how long this type of event historically impacted riders - up to 2 1/2 hours in this case. In addition, they provided information on alternate branches and stations that could accommodate riders and parking details at those stations.

The railroad also has an active twitter feed from which I received informative up-to-the-minute tweets from the railroad. Passengers tweeted 'boots on the ground' info. TV and radio stations the railroad communicates with also tweeted updates on a regular basis.

Using all of these sources of information I was able to wait at home with a hot cup of coffee instead of in my car or on the platform indefinitely.

At 7:21 AM the railroad updated the TrainTime app, sent text messages and tweeted that the line was back in action and the next train would arrive at 7:46 AM. Interestingly that is the train I take when I do not go in early.

I also took advantage of the TrainTime app 'Arrival Countdown' to determine exactly when the train would arrive. The LIRR also posts the minutes until arrival on station platform video boards, but accuracy can be enhanced in some instances.

In addition to text messaging, emails and Twitter the railroad also uses other social media platforms including Facebook, Instagram, YouTube, Snapchat and Flickr to communicate. I think all of this is cool for a railroad that dates back before Abraham Lincoln was President.

I finished my coffee and scooted out of my house. Jumped in my car and took my leisurely 12-minute drive to the railroad. The 7:46 AM was on time to the minute. Yippee!

I was surprised that the station was empty and not too many passengers boarded on the next few stations going west. Either they stayed home, went to another branch (that is a hassle) or drove into work. Perhaps, some of them would have been on my train but they may not use the digital communication channels the LIRR uses to alert passengers.

Although it was a challenging event for the railroad and inconvenienced riders I was delighted with the way they handled it end-to-end.

Of course, in our rapidly changing digital world there is many opportunities for the railroad to improve their customer digital delight. Recent trends give me confidence they will continue to improve.

In my day job as Director of Business Continuity for major companies, I have been through almost every type of natural and person-made disruptive event you can imagine. Communication is critical and in this case the LIRR aced it!

Update! As my book was going to press my faith in the LIRR was validated! MTA which runs the LIRR announced the beta release of the new MYmta app which consolidates three former apps. I am confident by the time you read this the production release will be available.

The new app contains accurate data directly form the MTA and includes the LIRR, subways, busses and Metro-North. It is packed with nice benefits such as real-time departure info, travel preferences, planned and unplanned service changes, personalized trip planning and what's nearby. They also stated that they are working on 'more features to come'.

Perfect timing LIRR and thanks for working hard to make all riders happy!

How Grandma Became an Unlikely Digital Success Story

GrandmaBetty.com
the Starting Point for Baby Boomers and Seniors

"Over 1,000 Info Sources and Products for Baby Boomers and Seniors as seen in The New York Times, Wall Street Journal, Fox News, BBC, CBS..."

GrandmaBetty's Categories:

- Health and Fitness
- Gardening
- Second Act Jobs/Careers
- Antiques and Memorabilia
- Other sites of interest
- Travel
- Entertainment
- Lifelong Learning Center
- Games
- Technology for Seniors
- Grandparenting
- Coffee
- Recommended Books
- Government
- Weather

    This digital entrepreneur's story is a great example that all it takes is a great idea, optimism, passion, a winning attitude, a great business model and the will to execute. Lean, agile, creative, innovative people such as this grandma can disrupt any industry. Age is not a factor nor is lack of experience.

    GrandmaBetty.com is one of those "you wouldn't believe it if you weren't there" stories. I interviewed Grandma Betty, who likes to be called GB, in her living room over non-stop courses of bagels and lox. After the second or third bagel, she forced more on me and told me I 'eat like a bird'.

    Interestingly, this is not the first business book GB has been featured in. She has appeared in a few. For example, Dan Pink, a New York Times best-selling business author featured GB as the first case study in his best seller, 'Free Agent Nation' and referred to her in subsequent interviews as one of the most fascinating people he has met.

    Dan wrote a great story but I know GB better, as she is my mother. I had a front row seat and saw the whole seemingly impossible digital success story unfold before my very eyes. I will let her describe the whirlwind events that occurred. Remember, it can happen to you!

Me - GB can you please take us through your digital journey and share any lessons learned?

GB - It would be my pleasure. What will provide the most value to your readers is if I begin with the press release we wrote at the time. It describes my predicament and what I did about it. Perhaps, it will inspire people and they can borrow bits and pieces for their own use. I only want them to be successful.

Here is the Golden Press Release that blew out Grandma's server! Please keep in mind we are not professional copywriters. We wrote it from the heart and, as you will see, it struck a chord.
For Immediate Release
Grandmother Strikes Back!!!
They all laughed when she sat down at the computer.
New York, NY
Less than 5 months ago, a 68 year old Grandmother was effectively put out to pasture when her company relocated and she was faced with a future of no income, no pension and having to bag groceries at minimum wage. In the 5 months since then she has made the most of a difficult predicament to develop a comprehensive Senior Portal on the Internet. This real-life Grandmother, her physical world name is Betty Fox - thus - GrandmaBetty.com, a youthful person who still wanted to contribute to society, sought out a way to add value to people's lives. Here is Fox's story in her own words:

A little after my `layoff' my son, the computer genius, made me aware of the Internet. He told me I did not have to become a computer genius to have the Net change my life. I did not know Internet from Shminternet but I have always loved to communicate with people, research things and to read. I soon realized that this is what the Internet is all about and that it is fast becoming the most important communications tool ever invented.

My son gave me one of his old computers and put something called a browser on it. It was slow and when he saw how much time I was spending online and how much more I wanted to do with the Internet he got me WebTV, which he connected to my television set. In minutes I was online and a whole new world opened up to me.

The next big problem was finding things on the Internet. I tried the various popular search engines aimed at the public. Oy, what a mess. On my first search attempt I was looking for a new recipe for peach pie. I went to one of the popular search engines and typed in the word 'peaches'. In a few seconds the search engine told me it found 85,870 pages with the word 'peaches' in them. I felt like I had just won the lottery. So, I started looking at the list of sites the search engine found. I nearly fell off the couch with disbelief...

There were thousands of pornographic sites with girls referring to themselves as 'peaches', thousands of sites with nothing to do with peaches at all, peach wholesalers in South America...on and on. I could have spent 6 months going through that list without finding a single peach pie recipe.

Next I tried a search for Seniors. I wanted to find sites for people like myself. Again, the result listed thousands of sites (805,750 to be exact). I took out my calculator and with my son's help figured out that if I spent 1 minute on each site it would take me 559 and a half days to visit each site and that's 24 hours a day and every day for a year and a half. I do not think that is a good investment of time. Many of the sites were by seniors in High School and seniors in College. Pornography sites galore, outdated information from 5 years ago, broken link after broken link to sites which no longer exist. And these results are on search sites being valued at billions of dollars. There had to be a better way.

Well, if you want it done, do it yourself. I decided then and there that I was going to devote my life to building the best Senior Web Site. The site would be for seniors anywhere in the world and the Internet would be our vehicle to communicate as a group and leverage our real-world knowledge. A knowledge base greater than that of any other demographic group.

GrandmaBetty.com is now who I am and what I do. I spend my full time visiting each and every site that goes on GrandmaBetty.com. Each category on GrandmaBetty.com makes sense to seniors. Each site you see on GrandmaBetty.com has been checked for quality and

relevance for active seniors. Sorry 'peaches the slut' please list your site elsewhere.
End of Press Release

GB - So we had our great Press Release but we did not have a list of anyone to send it to. We contacted PR Newswire, which is sort of like the Associated Press but less expensive. We found the cost of sending a national press release was $250. Marty and Joe each put in $125 and Joe emailed the release to PR Newswire for release at 7AM on July 18th. I am glad they were able to split the cost since I was unemployed/force-ably retired and had to watch my pennies.

You Go Girl!
The morning of July 18th was nerve wracking. I had no idea what was going to happen. I took two aspirins, which made me feel much better, and waited. I suppose my hope was that maybe one or two of the reporters who received the press release would come up to my site and send me a little note telling me they enjoyed it and to keep up the good work. Unfortunately, what happened that morning was nothing. No emails, no calls, no encouraging messages. Needless to say, I was disappointed. I called Joe and he told me not to worry as it sometimes takes time. He felt that someway, somehow, someone would find our little needle in a haystack. Was he ever right!

I woke up on the morning of July 19th and had messages from people who visited my site that night and took the time to email me. I started to realize that people from Arizona, California and New Mexico were using my site while I was sleeping. Some mentioned they heard of it from a small article in the Arizona Reporter.

The ensuing days brought hundreds of new visitors to my site and I was getting some amazing questions from visitors.

I thought the Arizona Reporter would be the end of the publicity from the press release, but again I was wrong. Suddenly I started getting emails and traffic from the Midwest and from the South. A newspaper in Wisconsin and a newspaper in Georgia both picked up my story and people were spreading the news by word-of-mouth. The number of visitors started growing dramatically.

The following day I received a call from a reporter at Newsday in New York. They had a circulation of 1,000,000. The reporter was Jamie Bernstein and he asked me who created the press release. I told him I did it from my heart with a bit of help from my two sons. He said it was great and he wanted to profile me in a Newsday article. I could not believe it, my first interview!

That interview opened the floodgates and people began pouring into GrandmaBetty.com. It tripled the number of daily visitors. Inquiries and suggestions began filling up my email box. I spent a good portion of each day personally answering each email. I felt anyone who took the time to visit my site and to contact me deserved my attention. This is true even today. It is one of the most important services I provide my wonderful visitors. Customer happiness makes all the difference. I advise many leaders and entrepreneurs to answer each inquiry quickly. This can make the difference between a sale and no sale.

Up-up and away! Fred Brock of The New York Times wrote an in-depth article that catapulted GrandmaBetty.com into the stratosphere. The story appeared in the Sunday edition of the Times, which is the most widely read issue of the week. In addition, and to my astonishment, it appeared in the popular Business Section of the paper. If that is not enough, the New York Times devoted three quarters of a page to me and my site!

The article in the New York Times came out on Sunday September 18th. When I opened my email program that morning, I already had over 839 messages from Saturday night and early Sunday morning. As the day progressed, the emails were coming in first one every five minutes, then by late afternoon they were coming in one every 15 seconds and by evening one every 5 seconds. We were in an avalanche of emails and there was no letting up! I hope you have the good fortune to experience this someday.

The Times article was generating inquiries from 50 States, Europe, Asia and Africa. People in countries such as Borneo (seriously) and Malaysia contacted me. It was thrilling. Viral marketing at its finest. Who would believe it? GrandmaBetty.com was now a global icon!

Joe, Marty and I were on the phone most of the day and watching in real-time as the avalanche of emails kept pouring in. Prior to launching our site Marty had configured my email program to produce an audible 'bing' sound whenever I received an email. He had heard that Jeff Bezos had done this when he developed Amazon.com and wanted an audible queue whenever an order was placed on Amazon.

We figured the emails would be 'few and far between', so I would not have to constantly check the computer for new emails, the sound would alert me. I encountered exactly the opposite problem. The onslaught of emails was causing the computer to 'bing' so fast I thought I was living in the middle of a shooting gallery in an arcade. We finally had to disable the 'bing', as I was getting 'binged' in the noggin every second of the day and night.

Tips from Grandma the Digital Entrepreneur:
- Treat your customers like gold. If they love what you are serving they will tell their friends and family. Stay active and listen to your customers. If you respect your customers your biggest worry will be how to handle hyper-interest and rapid growth.
- Always concentrate on improving your product or service. New technology offers new benefits. Don't live on your laurels. Stay ahead of the competition.
- If you only build it, they will not come. The reason so many web sites are not successful is that the owners somehow believe that if they simply build the site people are going to come surfing on by. Maybe it is the notion of the Internet being some sort of Information Superhighway that has influenced this thinking. They might imagine their site is like a McDonalds on the side of Route 66 with a big flashing sign and that people will magically pull off the superhighway and come on in to shop or buy a virtual hamburger. Sorry, it is not like that at all. Joe, Marty and I were realistic, as Joe was an expert on Internet marketing and we knew there were millions of sites and ours would have to stand out.
- If I can do it, you can too. Find a way to help people. Always provide value. That is what it is all about. Be your unique genuine self. Don't be afraid to be different. If we had positioned my site like every other site and a corporate voice as the face of the site, it would not have succeeded. None of the wonderful things that occurred

would have happened and you would not be reading my digital success story.

Just when I thought it could not get any better I received a call from People Magazine. I first thought it was a joke. I thought only celebrities and movie stars were featured in People. The reporter assured me it was no joke or mistake. They loved my story and wanted me. Incredibly I was featured in a full-page article. They wrote about my site and used a picture of me dribbling on a basketball court wearing my sons Air Jordan sneakers. It was ridiculously great.

Business professors contacted me letting me know I was an inspiration and part of their lectures. Students let me know I was an inspiration to them to launch a digital career.

Donnie and Marie asked me to be on their TV show to cook Chicken Soup.

Microsoft had conversations with Joe about involvement with WebTV.

I was featured in The National Inquirer and in newspapers and magazines on every continent except Antartica. I still have a glossy magazine article in Chinese with a picture of my site.

I was asked to be the Keynote Speaker at a National Customer Service Conference in Las Vegas. I said ok, they put me the cover of their beautiful glossy brochure and then I got cold feet, as I do not like to fly. Instead, my sons went to Vegas and I was on speaker-phone. Happily, the audience loved it!

Soon after that four companies wanted to acquire GrandmaBetty.com and me. The selection process was so much fun and so much pressure. My major concern was that whichever company we selected would benefit my members. Eventually we were absorbed into a larger company with nice people where the oldest person after me was 27 years old. I happily worked with them for many years.

What goes around comes around. I eventually got GrandmaBetty.com back. It is my love and passion. I maintain it daily to this day and always will.

Me - GB do you have any parting advice for my readers?
GB - Never give up. Have Fun. Work hard. Always do your best and I know you will find success.

About the Author

Marty Fox has achieved success as a digital business technologist and serial entrepreneur. He has developed successful business-to-business software systems, created unique digital assets enjoyed world-wide and authored thirteen popular business and technology books. He has also used digital techniques to overcome adversity and succeed.

Marty's 'lives' where business and technology intersect. His passion is demystifying complex technology and sharing how it can be used to disrupt industries, drive revenue, reduce expenses and improve the lives of many people.

Please join Marty on the road to success and happiness.

You can reach Marty by email:
Mfox@DigitalSuccessBook.com

Special FREE Bonus
Digital Success Newsletter

Digital transformation and disruption are quickly changing the world. It is important for you to understand and stay ahead of change.

As a free bonus to my book readers I would like to continue providing you with value after the book with a free subscription to my Digital Transformation Success Secrets Newsletter. Change does not stop and neither can you.

If you do not enjoy the newsletter you can cancel at any time.

Each issue is packed with:
- Late breaking digital business and career alerts, tips and techniques
- Digital software and hardware recommendations, reviews, free trials and product discounts
- New career success opportunities, success habits and possible pitfalls
- A comprehensive online listing and links to ALL resources in this book and many future digital companies and products as I discover them

To begin your free subscription today visit DigitalSuccessBook.com/newsletterbonus

A Special Request from Marty

Thank you for taking this digital journey with me. I hope you found it valuable. If you enjoyed Digital Transformation Success Secrets I would appreciate it if you would leave a rating and short review on Amazon. It is quick, easy and helps future readers and me. You can simply go to http://www.digitalsuccessbook.com and click the review the book link.

Thank you,
Marty

www.ingramcontent.com/pod-product-compliance
Lightning Source LLC
Chambersburg PA
CBHW030608220526
45463CB00004B/1220